AF873800

CRACKING A-LEVEL ORGANIC CHEMISTRY DEDUCTIVE QUESTIONS

Think Like A Detective!

CRACKING A-LEVEL ORGANIC CHEMISTRY DEDUCTIVE QUESTIONS

Think Like A Detective!

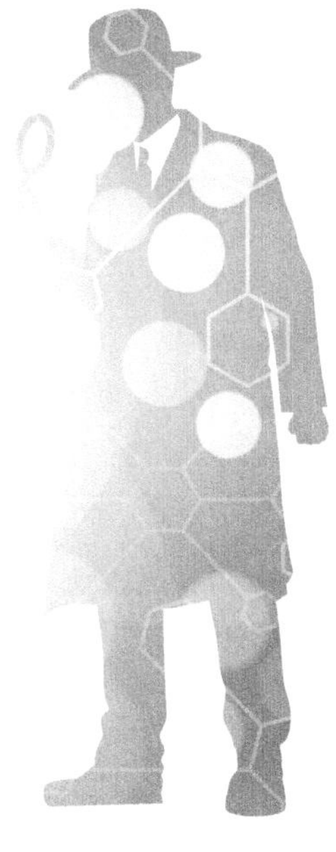

David Tan K.W. (MSc, NUS)
Leonard Pereira (PhD, NUS)

WS Education

NEW JERSEY • LONDON • SINGAPORE • BEIJING • SHANGHAI • HONG KONG • TAIPEI • CHENNAI • TOKYO

Published by

World Scientific Publishing Co. Pte. Ltd.
5 Toh Tuck Link, Singapore 596224
USA office: 27 Warren Street, Suite 401-402, Hackensack, NJ 07601
UK office: 57 Shelton Street, Covent Garden, London WC2H 9HE

British Library Cataloguing-in-Publication Data
A catalogue record for this book is available from the British Library.

CRACKING A-LEVEL ORGANIC CHEMISTRY DEDUCTIVE QUESTIONS
THINK LIKE A DETECTIVE!

Copyright © 2026 by World Scientific Publishing Co. Pte. Ltd.

All rights reserved. This book, or parts thereof, may not be reproduced in any form or by any means, electronic or mechanical, including photocopying, recording or any information storage and retrieval system now known or to be invented, without written permission from the publisher.

For photocopying of material in this volume, please pay a copying fee through the Copyright Clearance Center, Inc., 222 Rosewood Drive, Danvers, MA 01923, USA. In this case permission to photocopy is not required from the publisher.

ISBN 978-981-98-1820-4 (hardcover)
ISBN 978-981-98-1796-2 (paperback)
ISBN 978-981-98-1797-9 (ebook for institutions)
ISBN 978-981-98-1798-6 (ebook for individuals)

For any available supplementary material, please visit
https://www.worldscientific.com/worldscibooks/10.1142/14441#t=suppl

Typeset by Stallion Press
Email: enquiries@stallionpress.com

Printed in Singapore

Book Summary

Unlock the secrets of organic chemistry and turbocharge your grades with *Cracking A Level Organic Chemistry Deductive Questions: Think Like A Detective* by David Tan and Dr. Leonard Pereira. Step into the shoes of a molecular detective, inspired by Sherlock Holmes, and master the art of decoding complex structures with confidence. This proven guide features **81 exclusive structure elucidation questions** and equips you with foolproof strategies to tackle even the most perplexing problems.

Say goodbye to rote memorisation and hello to logical deduction, strategic thinking, and exam excellence.

Plus, get exclusive bonus content when you sign up at www.reviewston.com! Access a complete sample lesson, downloadable notes, and AI-marked MCQs with instant feedback — all designed to supercharge your learning.

Don't miss this opportunity to transform your organic chemistry journey. Grab your copy today and start thinking like a top scorer!

Table of Contents

About the Authors

David Tan

David Tan has been teaching O- and A-level Chemistry for over 20 years, during which he has helped thousands of students not only ace their exams but also fall in love with the subject. Many of his former students have gone on to become chemistry educators themselves — a testament to his inspirational and lasting impact.

David holds a Master of Science in Chemistry from the National University of Singapore (NUS), awarded in 2018, and brings deep academic insight to his teaching. His lessons are known for their clarity, intuition, and real-world relevance.

Outside the classroom, David is an avid reader and a competitive chess player. A former two-time national player for Singapore and currently rated 2044 by FIDE, he brings the same logical precision and strategic thinking from the chessboard into solving the toughest chemistry problems — especially in the field of **structure elucidation**, the focus of this book.

Dr. Leonard Pereira

Dr. Leonard Pereira earned his Ph.D. in Chemistry from the National University of Singapore (NUS) in 2004 and spent over a decade teaching at the university, where he gained a reputation for making advanced chemical concepts clear and approachable.

With a background that includes experience as a school teacher, Dr. Pereira brings a rare blend of academic depth and classroom understanding. He is especially aware of the real-world challenges students face when trying to master Chemistry, and now focuses his time on educational writing and mentoring.

This book, **Cracking A-Level Organic Chemistry Deductive Questions: Think Like a Detective**, marks the second collaboration between David and Dr. Pereira, following their bestselling title *A-Level Chemistry's Best Kept Secrets: What Top Students Know That You Don't*. That book transformed the way many students approached Chemistry. Now, in response to overwhelming student feedback, they've set their sights on what is arguably the most feared part of the syllabus — **organic chemistry structure elucidation**.

Preface

Dear Reader,

"Cher, organic chemistry is so tough!"
If you've ever found yourself uttering those words, you're certainly not alone. Many students find themselves overwhelmed by a tangle of reagents, conditions, and mechanisms — often resorting to massive mind maps, only to feel even more confused.

But what if you approached organic chemistry not as a memorization marathon, but as a mystery to be solved?

This book invites you to think like a detective — to channel your inner Sherlock Holmes. A-level organic chemistry, especially structure elucidation, can feel like a maze. Yet, with the right tools and mindset, you can navigate it with confidence and even excitement.

In these pages, we go beyond just presenting information. We'll show you how to apply unconventional but highly effective strategies: analyzing degrees of unsaturation, leveraging smartly constructed mind maps, and thinking in patterns. These techniques are designed not only to help you score marks but to build true understanding.

Yes, exams are timed and stressful. But true mastery comes from wrestling with problems, letting ideas percolate, and revisiting them with fresh insight — just like letting tea steep until the perfect brew. It's in this process that real learning happens.

To support your journey, I've created a companion website at www.reviewston.com, filled with bonus resources, practice questions, and walkthroughs that complement this book. (You may also discover a few videos there — I'm slowly building a visual library for learners like you.)

If you ever feel stuck or inspired, I'd be delighted to hear from you. You can reach me at david@reviewston.com — whether to ask a question, share an insight, or suggest how this book can be even better.

So, to all the curious minds — my fellow detectives of chemistry — let us step into the world of carbon clues and molecular mysteries. With curiosity as our guide and logic as our lantern, we shall unravel even the most enigmatic of structures.

Acknowledgements

I am deeply grateful to my students — past and present — whose curiosity, resilience, and relentless questions have been the heartbeat of this book. Your enthusiasm has not only sustained my passion for teaching but also constantly reminded me that true education is a two-way journey. Our lively discussions and shared moments of discovery have shaped this book in more ways than you know. You have pushed me to think deeper, teach better, and never settle — for that, I am truly indebted.

My heartfelt appreciation goes to my teacher, mentor, and dear co-author, Dr. Leonard Pereira. This marks our second collaboration, and I continue to be enriched by his intellect, humility, and unwavering generosity. His mentorship has left an indelible mark on my growth — not only as an educator but as a person. He has been a guiding light and a fatherly figure, and words cannot fully express my gratitude.

I am also sincerely thankful to the wonderful team at World Scientific Publishing Co. Pte Ltd, whose dedication and attention to detail have helped bring this vision to life. A very special thanks to Sandhya Devi, our editor, whose gentle encouragement and patience gave me the space to write with clarity and care. Her support has been instrumental, and I am genuinely thankful for her belief in this project.

As I reflect on this journey, I wish to honor the memory of my late father, whose quiet strength and unshakable faith in me continue to guide my every step. To my beloved mother — the most selfless and steadfast supporter I have ever known — your unconditional love has been my anchor, and I owe everything to you.

It's humbling to realize that it has been seven years since my first A-Level Chemistry book, *A-Level Chemistry's Best Kept Secrets: What Top Students Know That You Don't*, was published. Since then, my ideas have evolved, and the call from students for a book focused entirely on organic chemistry has grown louder. Many of you wrote to me, shared your struggles, and encouraged me to create something that could make a real difference. I heard you.

This book, *Cracking A-Level Organic Chemistry Deductive Questions: Think Like a Detective*, is for you. It is not merely a collection of questions — it is an invitation to think differently, to approach organic chemistry with curiosity, imagination, and confidence. Every example has been lovingly crafted, often tested in real classrooms, and refined with feedback from the very students who inspired it. There is humor, there is challenge, and above all, there is heart.

To my students — my ever-patient "guinea pigs" — thank you for walking this path with me. You have been my greatest teachers.

To everyone who believed in this journey — your support has been my greatest honor. If this book helps even one student see organic chemistry not as a wall to fear, but a puzzle to enjoy, then every hour spent writing it has been more than worthwhile.

With deepest gratitude,
Tan Kien Wei David (M.Sc, NUS)
Singapore, 3 May 2025

Introduction

Organic chemistry can feel overwhelming — a vast jungle of reactions, mechanisms, reagents, and mind maps. But what if you had a map? A way to *see* how everything connects?

That's exactly what this book — and its companion platform — aim to provide.

One of the most powerful tools I use in teaching is the **flowchart**, or what some might call a mind map — but not one you simply memorise. These are designed to help you develop a **macro perspective**: to see how different homologous series and functional groups — **alkanes, alkenes, arenes, alcohols, alkyl halides, carbonyl compounds, carboxylic acid derivatives, and nitrogen compounds** — are all interconnected.

Ask yourself:

- Can I convert an **alkane to an arene**?
- Can I go from **alkyl halide to a nitrogen compound**?
- More importantly, can I work **backwards** — from a nitrogen compound to any of the earlier functional groups?

By moving back and forth between transformations, you're not just learning reactions — you're building **dendrites** in your brain, forming solid mental bridges that deepen understanding. This kind of **bidirectional, relational thinking** is what allows students to thrive in organic chemistry, even under exam pressure.

This book focuses especially on **structure elucidation** — one of the most intellectually stimulating and challenging areas of the syllabus. You'll find **81 carefully curated and original questions**, organised into three levels of complexity:

1. **C, H, O** – The backbone of all organic molecules
2. **C, H, O, X** – Introducing halogens
3. **C, H, O, X, N** – Adding nitrogen and higher-level functionality

Many students have commented that some questions seem almost identical. And they're not wrong — this is **intentional**. In fact, some have even asked me, "Cher, what's the difference between this question and that one?" What I'm trying to highlight is this: **subtle changes in**

phrasing, sequence, or chemical tests can drastically change the final answer. Being attuned to these nuances is a skill — and it's the very skill that **sets top scorers apart**.

Isn't that exactly how **Sherlock Holmes** operates? He doesn't rely on sweeping generalisations — he solves mysteries by noticing the fine details others dismiss. In the same way, thinking like a detective means training your eye to catch the tiniest clues — because in organic chemistry, they make all the difference.

In my own tuition classes, I use my first book, *A-Level Chemistry's Best Kept Secrets: What Top Students Know That You Don't*, as a structured set of notes. This current volume builds on that foundation, focusing deeply on solving real, meaningful problems — and guiding students through a **process**, not just an answer.

To support students further, I've developed **reviewston.com**, a dedicated **Learning Management System (LMS)**. It's more than just a website — it's a digital classroom where students can:

- Access **free sample lessons** and **curated notes**
- Attempt **AI-marked MCQs** and get **instant feedback**
- Track progress, view explanations, and identify areas for improvement
- **Message me directly through the internal system** for help or guidance

Currently, selected materials are **free**, and the full platform is open to all my tuition students. But this is just the beginning.

I'm working on a much bigger, long-term vision — one that supports not only **A-Level Chemistry**, but also **A-Level Physics, Biology, and Mathematics**, alongside **O-Level Chemistry, Physics, and Biology**. This is a **noble and ambitious project** that will take time, care, and dedication — but I believe it can change how students learn, grow, and succeed.

In the future, I may introduce optional subscription features. However, my goal remains clear: to keep learning **accessible, affordable**, and **purpose-driven**. I intend to donate **80% of all profits** to educational charities, because education should uplift not just individuals — but entire communities.

Whether you're studying independently, attending tuition, or preparing for major exams, this book and its online companion are designed to help you:

- **See organic chemistry with clarity**
- **Solve with logic and confidence**
- And most of all, **think like a detective**

Let’s begin this journey together — one clue, one structure, one insight at a time.

Part 1

The Investigator's Toolkit

Essential tools, concepts, and mental frameworks to prepare for structure elucidation.

1 Organic Chemistry Isn't Complicated — It Just Pretends to Be

"Just as the frost conceals the landscape beneath, so too does complexity shroud the beauty of organic compounds — only through diligent inquiry can we unearth their hidden truths." — **DT**

Organic chemistry used to feel like a maze to me. So many reactions to remember. And just when I thought I had everything under control, some professor would casually toss in a rare, newly discovered reaction that threw me off completely.

But one day, a professor said something that changed everything:

> **"Organic chemistry is indeed complicated with all its reactions, but at its core, every reaction is either polar or non-polar."**

That was my **Eureka!** moment.

Suddenly, the chaos started making sense. Polar reactions? They involve **nucleophiles** and **electrophiles** — that's right, our old friends δ^- and δ^+. Non-polar reactions? They're all about **free radicals**.

And here's the kicker: at **A-Level**, 99% of what you need to know are **polar reactions**. The only exception? **Free radical substitution** — and that lives in the very first topic you learn: **alkanes**.

💡 Breaking Down Reaction Types

Let's simplify this even more.

There are only **three fundamental reaction types** in organic chemistry:

- **Addition** – you're adding something
- **Elimination** – removing part of a molecule
- **Substitution** – swapping one group for another

Now combine those with the **three main reacting species**:

- **Nucleophiles**
- **Electrophiles**
- **Radicals**

If you love permutations, you'll appreciate this little matrix:

Species ↓ / Type →	Addition	Substitution	Elimination
Nucleophile	☑ Nucleophilic addition	☑ Nucleophilic substitution	✖
Electrophile	☑ Electrophilic addition	☑ Electrophilic substitution	✖
Radical	☑ Free radical addition (polymerisation)	☑ Free radical substitution	✖

So really, we're only left with **six core polar/non-polar reaction types** to focus on:

- **Free radical substitution** (alkanes)
- **Free radical addition** (polymerisation – not a big exam focus)
- **Electrophilic addition** (alkenes)
- **Electrophilic substitution** (arenes)
- **Nucleophilic addition** (carbonyls)
- **Nucleophilic substitution** (halogenoalkanes, acid derivatives)

No need to memorise hundreds of mechanisms. Just recognise the species involved, know your reaction type, and the rest falls into place.

🧪 What About Fancy-Sounding Reactions?

Let's decode some of those intimidating terms:

- **Hydrolysis of RX** → That's just **nucleophilic substitution**.
- **Hydrolysis of esters or amides** → A classic example of **nucleophilic acyl substitution**.
- **Hydrolysis of nitriles** → Starts with **nucleophilic addition** to the C≡N bond, forming an amide intermediate, followed by **nucleophilic acyl substitution** to yield a carboxylic acid.
- **Condensation reactions?** Think of them as **addition followed by elimination** — especially when forming esters or amides.
- **Redox reactions** (e.g., oxidation of alcohols)? These sit **outside the polar/non-polar framework**, but they're conceptually simple — it's just **electron transfer**, not detailed mechanism work.

🔍 Organic Chemistry's Secret

Once you strip away the jargon, most organic reactions fall under just **a few categories**. The rest? They're just **variations on a theme** — fancy costumes over a simple skeleton.

It's like discovering organic chemistry's **secret recipe**:

- A small set of reacting species
- A handful of reaction types
- A few recurring mechanisms

And with this knowledge, you're no longer a confused student — you're the detective, spotting clues, decoding mechanisms, and solving molecular mysteries with the sharp reasoning of **Sherlock Holmes** in his study at **221B Baker Street**, surrounded by clues only the observant would notice.

---

2 Electrophiles & Nucleophiles — The Polar Duo

"In every bond, there lies a whisper of polarity—an unseen force echoing through the ages, reminding us that even in simplicity, complexity lurks just beneath the surface."
— **DT**

2.1 Polar vs. Non-Polar Reactions

Building on Chapter 1, recall that **every** organic reaction is either:

- **Polar** (involving electrophiles and nucleophiles)
- **Non-polar** (involving radicals)

At A-Level, we focus almost exclusively on **polar** reactions—except for **free-radical substitution** in alkanes.

2.2 Recap of Reacting Species

- **Radical**: An atom, molecule, or ion with an unpaired valence electron.
- **Nucleophile** (Nu^- or Nu): An electron-pair donor.
  - *Examples:* NH_3, CH_3OH, OH^-, CN^-, CH_3^-
- **Electrophile** (E^+ or E): An electron-pair acceptor.
  - *Examples:* CH_3^+, NO_2^+, $AlCl_3$, PCl_3

Rule of thumb: In a polar reaction, the **nucleophile always attacks** the **electrophile**. Mix them up and you'll mix up your product.

2.3 Why This Matters for Structure Elucidation

When you draw a mechanism or propose a product—say, adding CO_2 to a Grignard reagent—you're really asking, "Which is Nu, and which is E^+?" That decision tells you exactly where bonds form or break, steering you to the correct structure.

2.4 Functional-Group Survey

Below is how each major class behaves in terms of polarity. Simple **structural sketches** are included—no mechanism arrows, just the essentials you need at a glance.

- **Alkanes**
    - *Reactivity:* Non-polar
    - *Key reaction:* Free-radical substitution
    - *Why:* C–H bonds are essentially non-polar; radicals mediate the process.
- **Alkenes**
    - *Reactivity:* Polar nucleophiles
    - *Key reaction:* Electrophilic addition (the π bond donates to E^+)
    - *Why:* π-electrons are high-energy and ready to attack an electrophile.

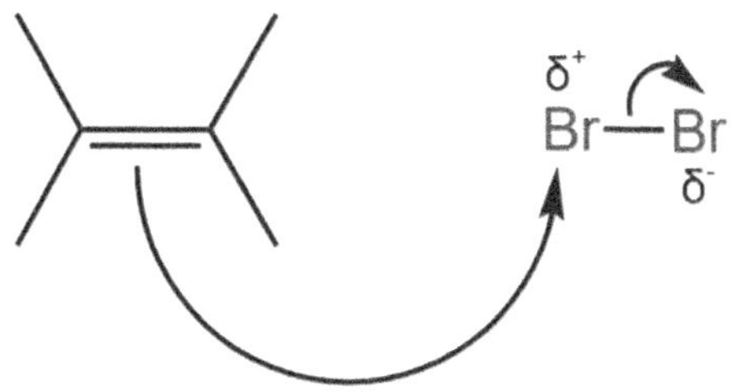

- **Arenes**
  - *Reactivity:* Polar nucleophiles
  - *Key reaction:* Electrophilic substitution (requires a strong E^{+})

    Generation of Electrophile: $AlBr_3 + Br_2 \longrightarrow AlBr_4^- + Br^+$

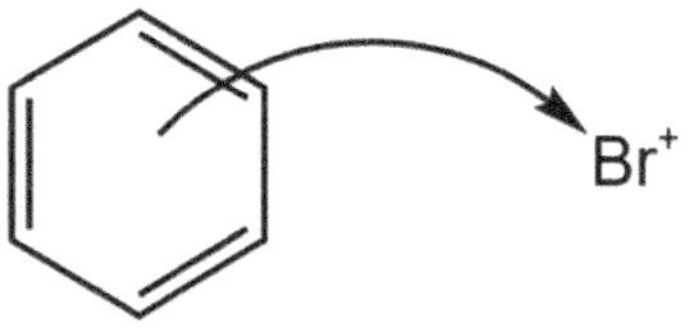

  - *Why:* Aromaticity must be preserved—addition would destroy resonance.

- **Hydroxyl Compounds (R–OH)**
  - *Reactivity:* Polar nucleophiles
  - *Reactions:*
    - Substitution with PCl_3 (–OH → –Cl)
    - Addition to carbonyls (hemiacetal formation)

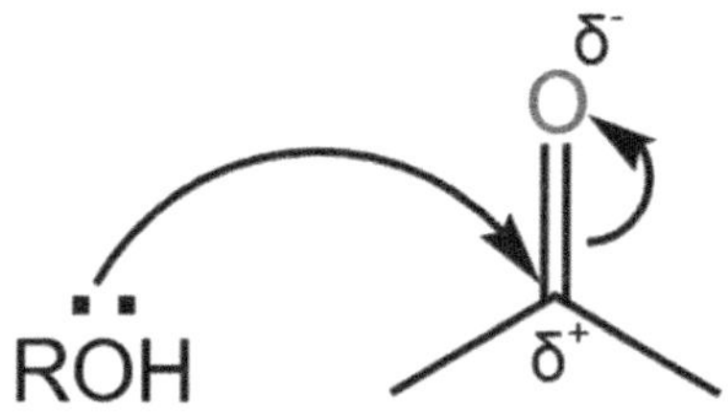

  - *Why:* Lone pairs on O make –OH a classic Nu.

- **Halogenoalkanes (R–X)**
  - *Reactivity:* Polar electrophiles
  - *Key reaction:* Nucleophilic substitution (S_N1 or S_N2)
  - *Why:* C–X bond is polarized ($C^{\delta+}$, $X^{\delta-}$), inviting Nu to attack.

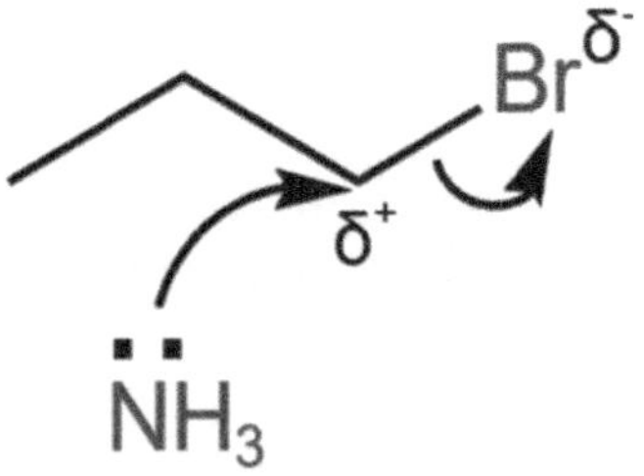

- **Carbonyls (aldehydes & ketones)**
  - *Reactivity:* Polar electrophiles
  - *Key test:* 2,4-DNPH
  - *Key reaction:* Nucleophilic addition to C=O
  - *Why:* C (δ^+) is attacked by Nu, π bond (weaker) breaks first.

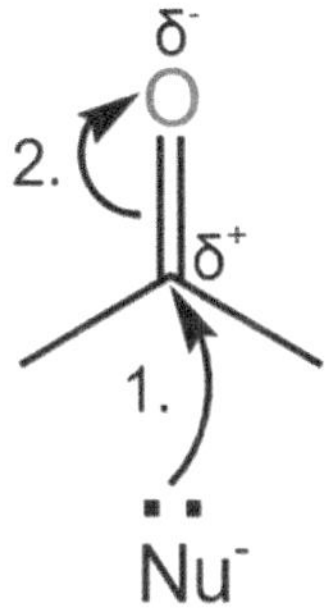

1. Nucleophile attacks electron deficient carbonyl C atom
2. Breaking of pi bond; O atom gains a lone pair

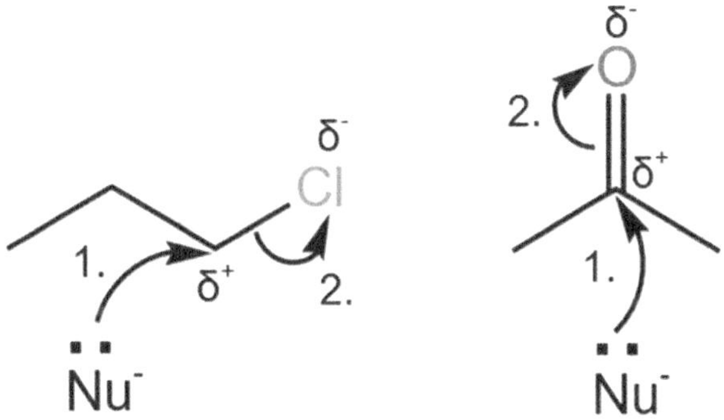

Note the similarities

- **Carboxylic Acid Derivatives**
    - *Reactivity:* Polar electrophiles
    - *Key reaction:* Nucleophilic acyl substitution
    - *Why:* Good leaving groups (–Cl, –OR, $–NH_2$) enable substitution.

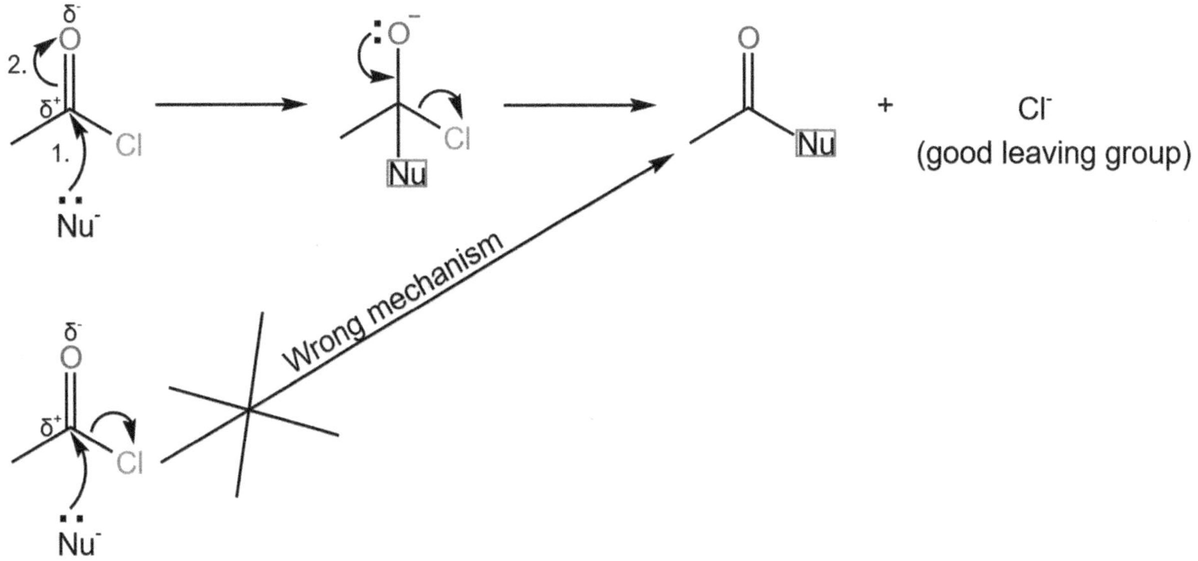

2.5 Decoding "Fancy" Reactions

- **Hydrolysis of R–X → Nucleophilic substitution**
- **Hydrolysis of esters/amides → Nucleophilic acyl substitution**
- **Hydrolysis of nitriles → Nucleophilic addition** to C≡N → amide → **nucleophilic acyl substitution** to acid
- **Condensation → Addition** then **elimination** (e.g., ester formation)
- **Redox reactions** (e.g., alcohol → ketone) → simple **electron transfer**, outside the polar-nonpolar framework

---

2.6 Key Takeaway

Organic reactions only wear many names; they all follow **addition**, **elimination**, or **substitution**, and involve just three players—**Nu, E$^+$, or radicals**. Learn to spot the pattern, and complexity melts away.

Onward to Chapter 3: *The Five Elusive Assassins: Meet Your Nucleophile Hit Squad* — a deep dive into the most common nucleophiles, and how to use them like the sleuth you are.

3 The Five Elusive Assassins: Meet Your Nucleophile Hit Squad

"To navigate the labyrinthine pathways of organic synthesis, one must embrace these five elusive assassins; for they are not mere reactants but rather architects of change, transforming mere substrates into masterpieces of molecular artistry."

— **DT**

3.1 From Polarity to Power

In Chapter 2 ("Electrophiles & Nucleophiles — The Polar Duo") we learned that every polar reaction pairs an **electron-pair donor** (nucleophile) with an **electron-pair acceptor** (electrophile). Now it's time to recognize the nucleophiles themselves—and to distinguish **nucleophilicity** from **basicity**:

- **Basicity** measures how unstable a negative charge is (how eager it is to grab a proton).
- **Nucleophilicity** measures how easily an electron cloud distorts to form a new bond (i.e. how polarizable the species is).

Basicity is how badly a species wants a proton—think of it as "proton hunger." **Nucleophilicity** is how easily it can attack another atom to form a bond—think of it as "attack skill."

Often a good base is also a good nucleophile (e.g., OH^-). **But** because larger atoms have "fluffier" electron clouds that can wiggle into place more easily, they can sometimes be *better* attackers even if they're *weaker* bases.

- In water, for example, iodide (I^-) is a poorer base than fluoride (F^-) but a *stronger* nucleophile—its electrons aren't held as tightly by the solvent, so it can swoop in on electrophiles more readily.

3.2 The Five Assassins' Roster

We group all nucleophiles into **five "assassins"**—each with unique stealth, strength, and selectivity:

1. **H-Assassin** – Hydride donors ($LiAlH_4$ & $NaBH_4$)
2. **O-Assassin** – Oxygen-based Nu's (OH^-, RO^-, R–OH, phenol)
3. **S-Assassin** – Sulfur-based Nu's (thiols & RS^-)
4. **N-Assassin** – Nitrogen-based Nu's (amines, hydrazine, aromatic amines)
5. **C-Assassin** – Carbon-based Nu's (Grignard reagents, Gilman reagents, Wittig ylides)

3.2.1 H-Assassin: The Hidden Power of Hydride

- **Reagents:** $LiAlH_4$ (strong) and $NaBH_4$ (milder)
- **Why they work:** Al or B delivers H^-; larger Al center is more polarizable than B, so $LiAlH_4$ reduces carboxylic acids & derivatives, whereas $NaBH_4$ only tackles ketones and aldehydes.
- **Selectivity:**
  - Carbonyl's C-atom (δ^+) is highly electrophilic → hydrides attack easily.
  - R–X's C-atom is weakly electrophilic → hydrides won't touch halogenoalkanes.

Halogenoalkane

Carbonyl, carboxylic acid & derivatives

Oxygen is more electronegative than chlorine

3.2.2 O-Assassin: The Electronegative Warriors

- **Anionic Nu's:** OH^-, RO^- (alkoxides)
    - **Basicity order:** $H^- > RO^- > OH^-$
    - **Nucleophilicity order:** $RO^- > OH^- > H^-$
- **Neutral Nu's:** R–OH (alcohols), phenol
    - Alcohols attack carbonyls (hemiacetal formation) with acid catalysis.

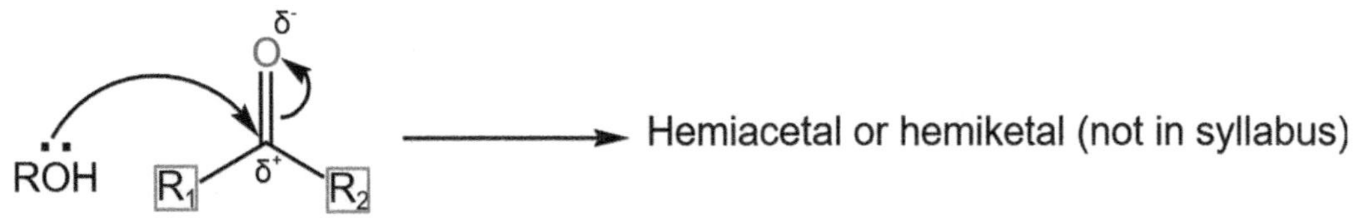

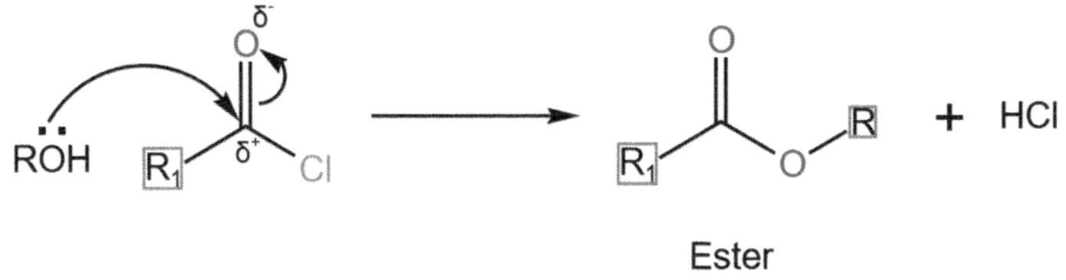

    - **Phenol's** oxygen lone pair is delocalized into the aromatic ring, making it a poor nucleophile at oxygen; instead, phenol prefers to participate in **electrophilic aromatic substitution** on the ring.

- **Key takeaway:** Alkoxides excel at S_N2; hydroxide is your go-to base; alcohols need catalysts.

3.2.3 S-Assassin: The Polarizable Powerhouse

- **Reagents:** Thiols (R–SH), sulfides (RS^-)
- **Why they shine:** Sulfur's larger, softer electron cloud makes S-Nu's even stronger than O-Nu's.
- **Use cases:** Thioether formation, thiol additions to C=O or acyl chlorides.

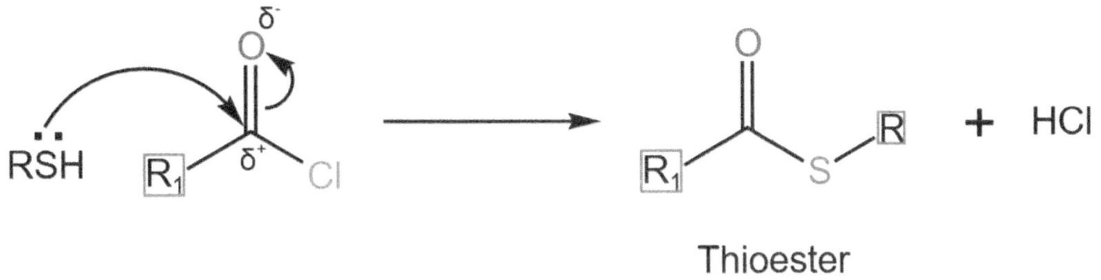

- *FYI:* Rare in A-Level, but watch for them in extension or research problems.

3.2.4 N-Assassin: The Lone-Pair Strategists

- **Reagents:** NH_3, R–NH_2, hydrazine (NH_2NH_2), aromatic amines
- **Dual role:** N-Nu's are also bases—direct amide formation with R–COOH gives acid–base, not substitution.

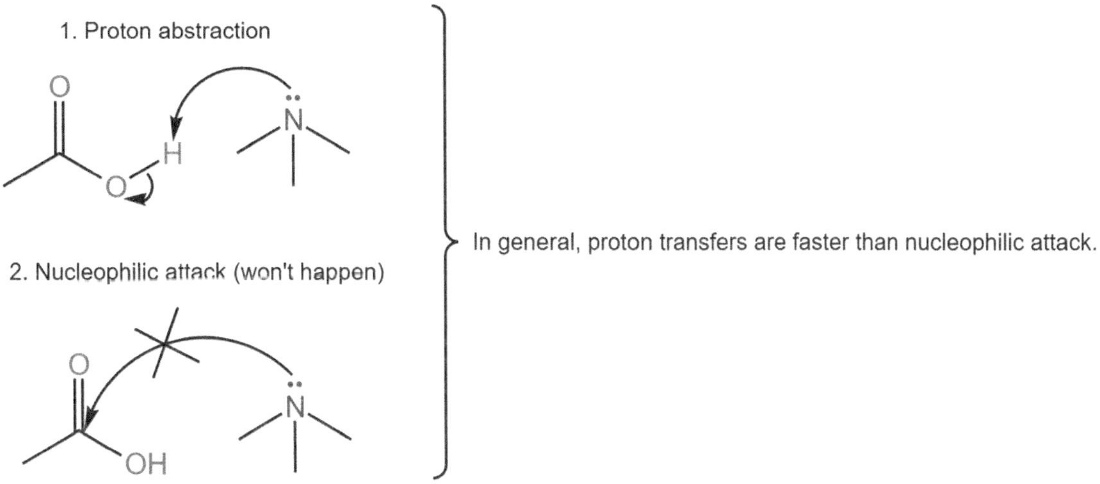

- **Tricks of the trade:**
  - Use acyl chlorides or anhydrides for amide synthesis.

1. No proton abstraction

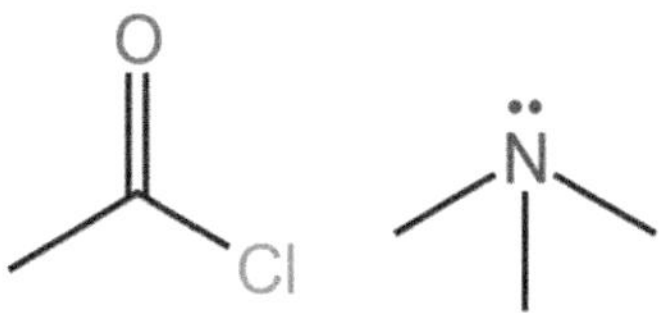

2. Nucleophilic attack happens instead

  - Hydrazine + C=O → imine (basis of 2,4-DNPH test).

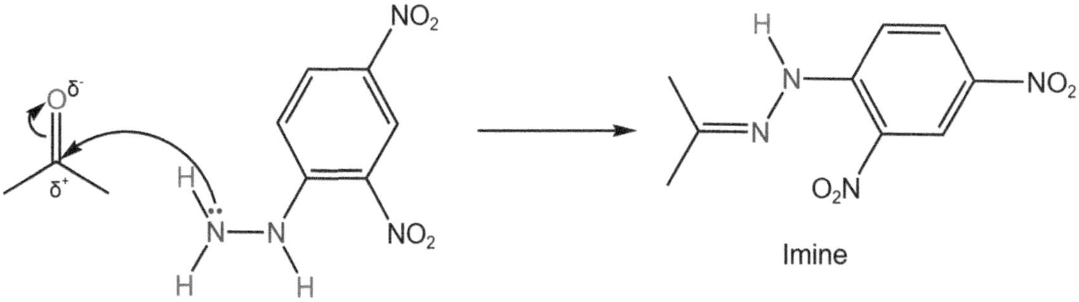

(A variant of hydrazine, N_2H_4)

  - Aromatic amines: lone pair delocalized—less nucleophilic than aliphatic amines.

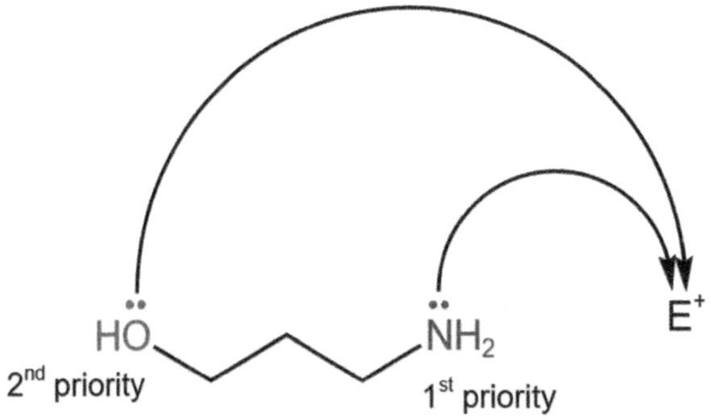

3.2.5 C-Assassin: The Carbon Chain Extenders

- **Reagents:** Grignards (R–MgX), Gilman reagents (R_2CuLi), Wittig ylides
- **Grignard essentials:**
  - Formed in dry ether; C–X polarity reverses (C becomes δ^-).

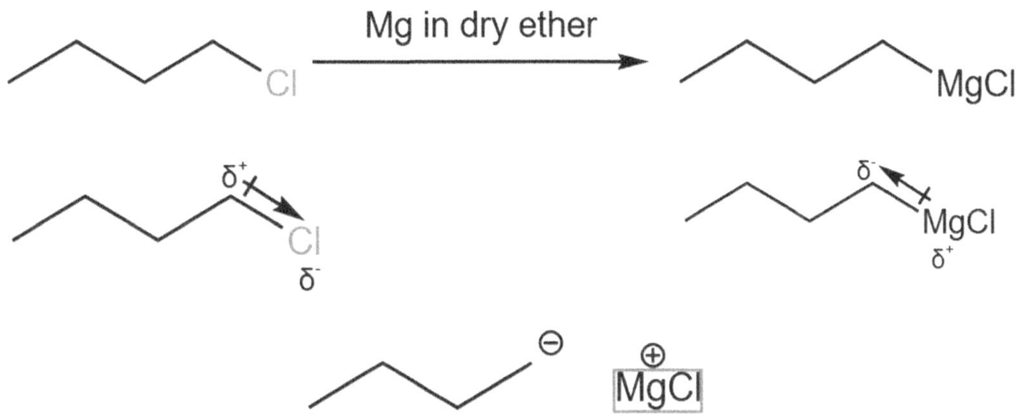

  - Reacts with CO_2 → carboxylic acids; C=O → alcohols; protic sources → alkanes.

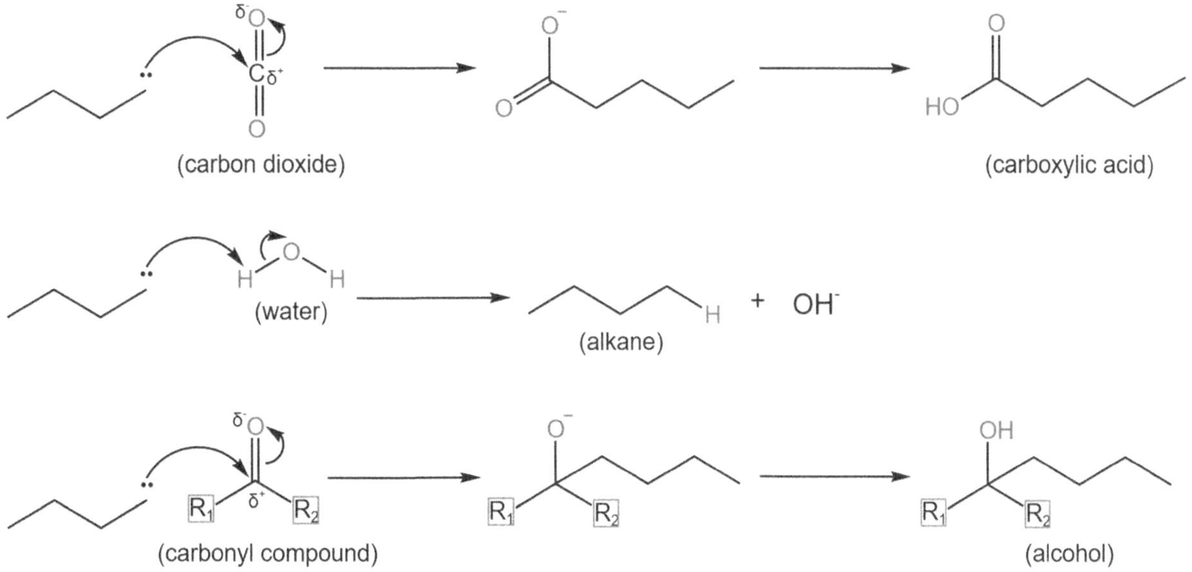

  - Must be water-free—extremely basic.

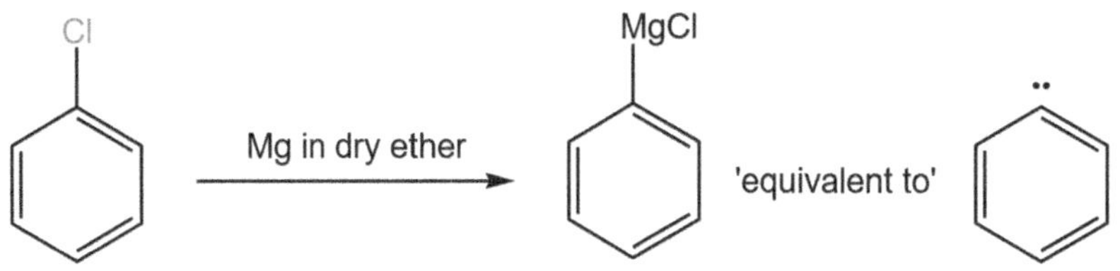

Example:

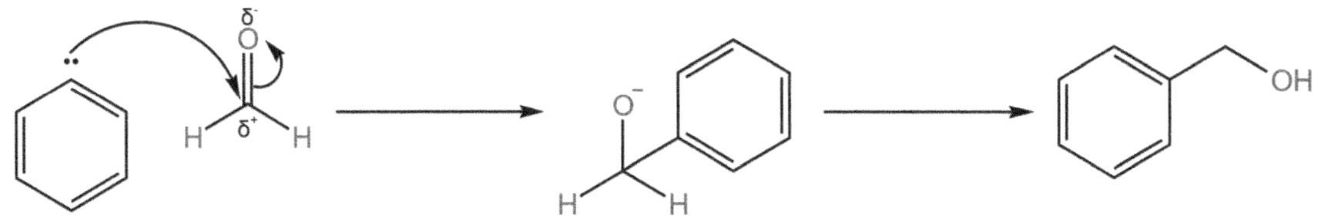

- **Gilman reagents:** Lithium-based coupling agents—less common at A-Level.
- **Wittig ylides:** Convert C=O to C=C—elegant carbon elongation.

3.3 Field Manual: Putting It Into Practice

1. **Spot the Nu:** Locate lone pairs or negative charges in your reactants.
2. **Rank the assassin:** Consider polarizability over raw basicity when choosing your Nu.
3. **Select your weapon:** Hydride for reductions, Grignard for chain building, etc.
4. **Execute the attack:** Nu attacks E^+—bonds form, leaving groups depart, and you claim victory.

Up next (Chapter 4): **The Arrow Detective — Tracking Every Bond Move** — because every hit squad needs a clear map of their operation.

4 The Arrow Detective — Tracking Every Bond Move

"To decipher the arrow code is to unlock a treasure chest of molecular mysteries; it is within these symbols that the alchemy of creation unfolds before our very eyes." — **DT**

4.1 Spot Your Players: Nu^- vs. E^+

Before you draw a single arrow, **identify your nucleophile** (electron-rich: lone pair or negative charge) and **electrophile** (electron-poor: δ^+ or full positive).

- **Who has electrons to give?** → That's **Nu^-**.
- **Who needs electrons?** → That's **E^+**.

Clear identification of Nu^- attacking E^+ is your first step to any mechanism.

4.2 Why Master Arrow-Pushing?

Curly arrows are the language of mechanisms. They track **electron flow**, revealing exactly how bonds break and form. Even in structure-elucidation puzzles, knowing which arrows to draw lets you predict products with confidence.

For example, when bromobenzene reacts with magnesium in dry ether, it forms a Grignard reagent, which can then react with benzophenone. If you understand the fundamentals of electron flow, you can easily predict the resulting product.

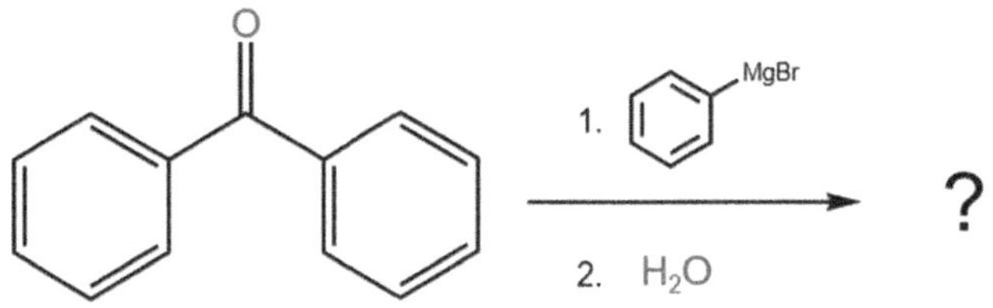

'Rough' mechanism:

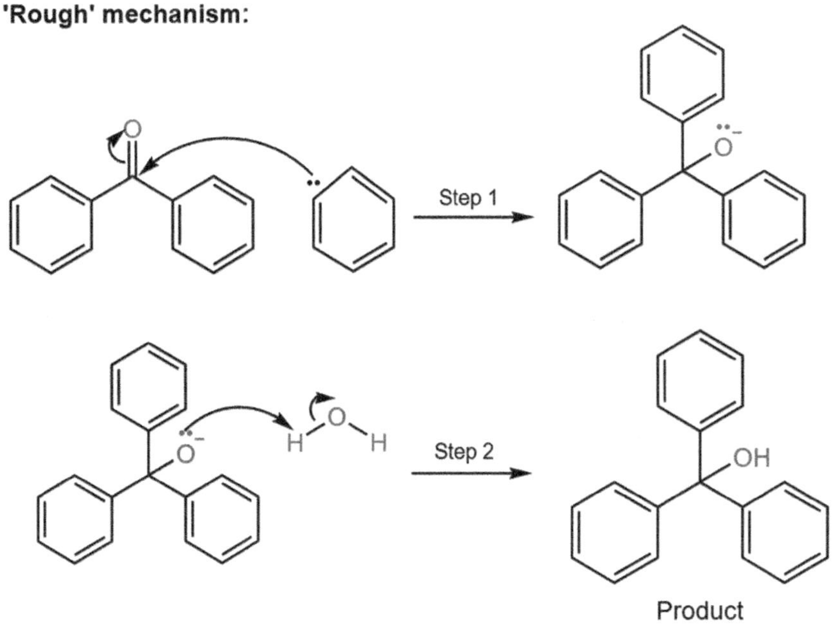

4.3 The Five Arrow Types

1. **Reaction Arrow (→)**
   Shows reactants → products.

2. **Equilibrium Arrow (⇌)**
   Indicates reversibility.

3. **Resonance Arrow (↔)**
   Links resonance forms—a hybrid exists, atoms stay put.

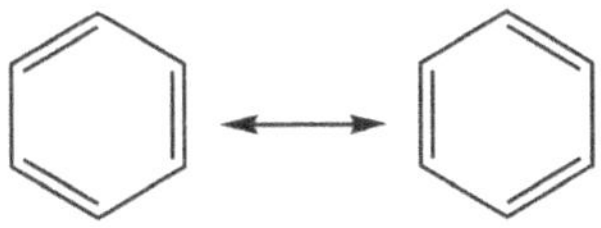

4. **Full-Headed Curved Arrow**
   Electron-pair movement (polar reactions):

   ○ From lone pair/bond **tail** → atom or bond **head**.

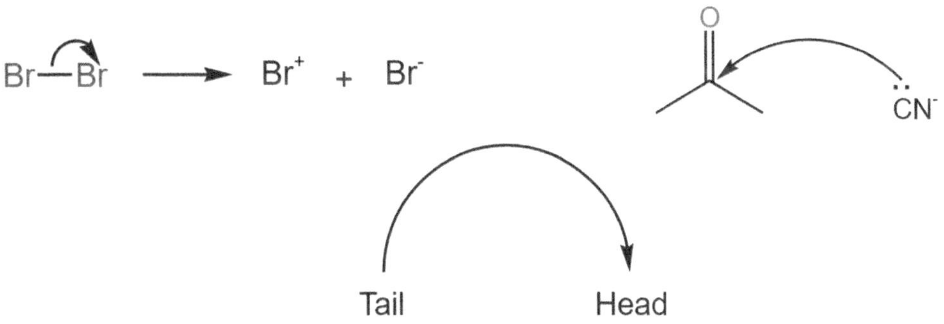

5. **Fishhook (Half-Headed) Arrow**
   Single-electron movement (radical, non-polar reactions).

Pro tip: Never start an arrow from a positive (or negative) charge—it's electrons you're tracking, not empty spaces.

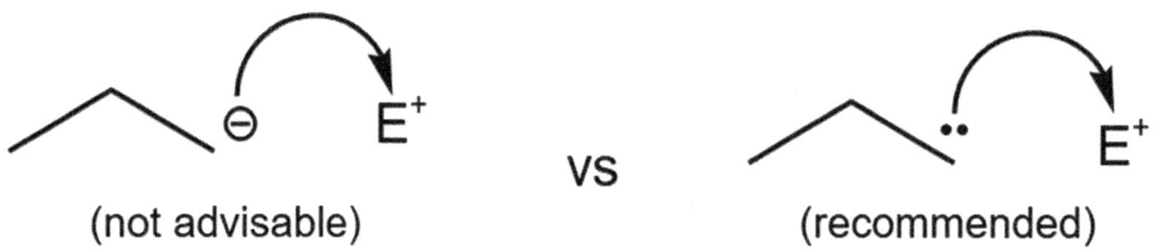

4.4 The Four Core "Moves"

All mechanisms—polar or radical—are built from these four steps:

1. **Nucleophilic Attack**
   - **Nu⁻ → E⁺** (tail from Nu's electron pair).

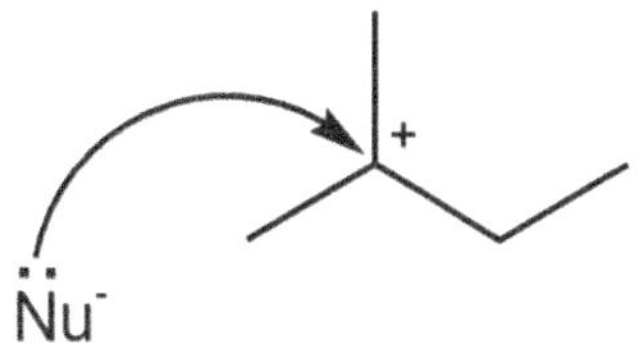

   - If a molecule has multiple Nu sites, the **larger or less electronegative** atom often strikes first.

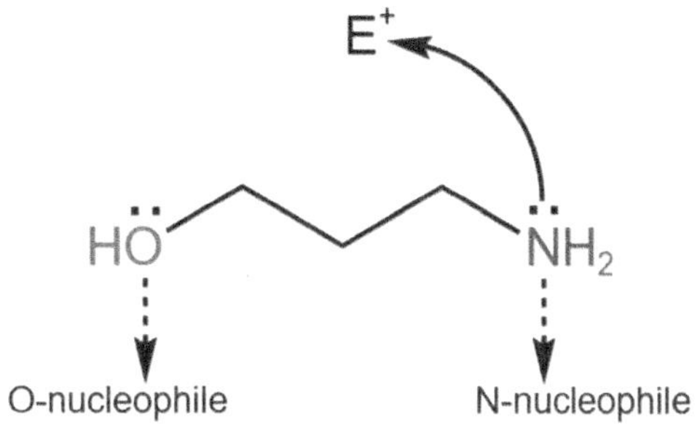

2. **Proton Transfer**
   - **Deprotonation:** Base's lone pair → H⁺.

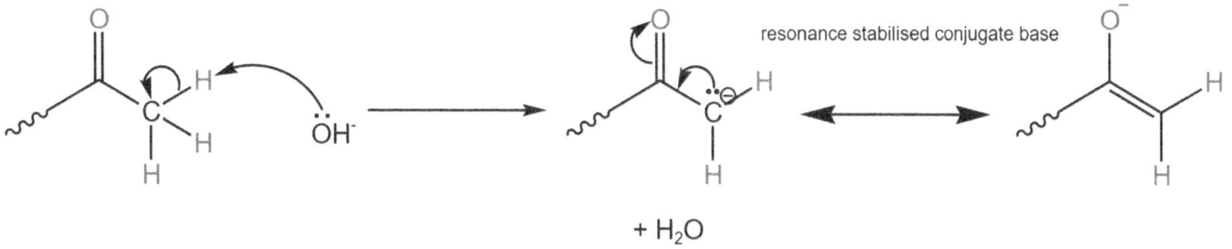

   - **Protonation:** Acid's H⁺ → lone pair or π-bond.

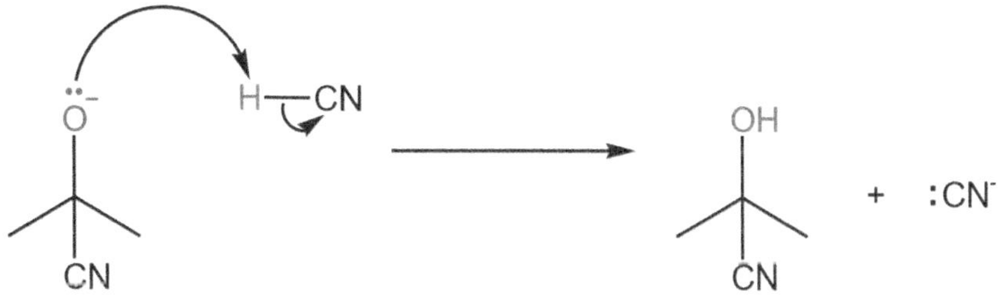

   - Stability of the conjugate base (resonance, inductive effects) dictates ease.

3. **Leaving-Group Departure**

   - Bond → LG, forming LG⁻.

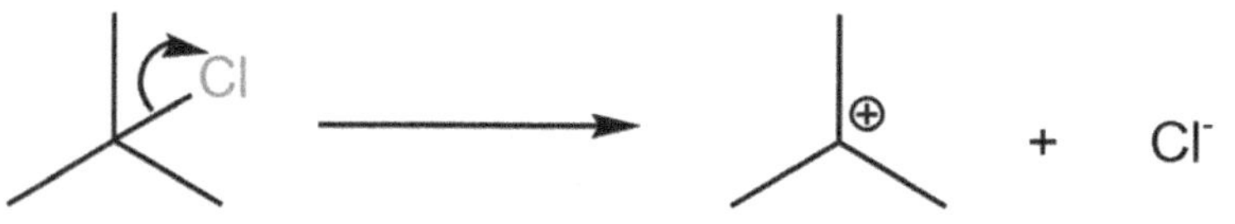

   - Good LGs (Cl^-, Br^-, TsO^-) depart readily.

4. **Rearrangement**

   - **Hydride** or **alkyl shift** to form a more stable carbocation or radical.

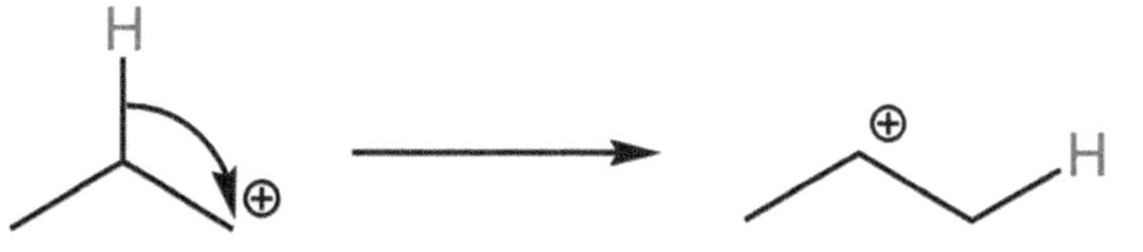

Primary carbocation (LESS stable) — Secondary carbocation (MORE stable)

   - *Example:* Unexpected Friedel–Crafts product due to hydride shift.

Generation of Electrophile: $AlCl_3 + CH_3CH_2CH_2Cl \longrightarrow AlCl_4^- + CH_3CH_2CH_2^+$

Primary carbocation (LESS stable)

Product: what we expect to get

In reality...

Primary carbocation (LESS stable) — Secondary carbocation (MORE stable)

followed by...

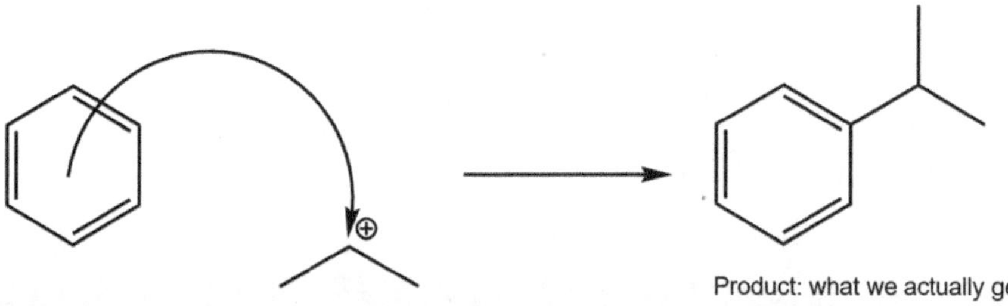

Product: what we actually get

4.5 Worked-Problem Spotlight

Your mission: Identify the mistake(s) in the mechanism drawing for each example.

(a)
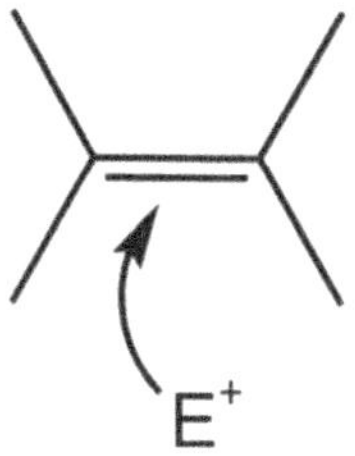

(b)
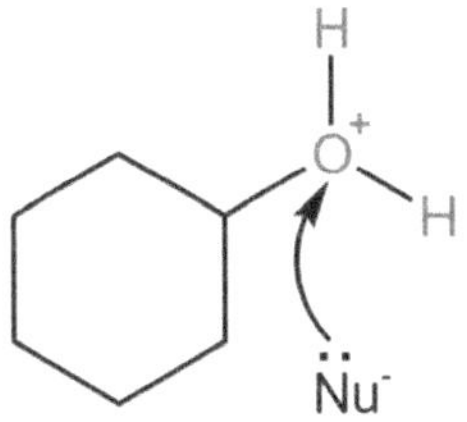

(c)
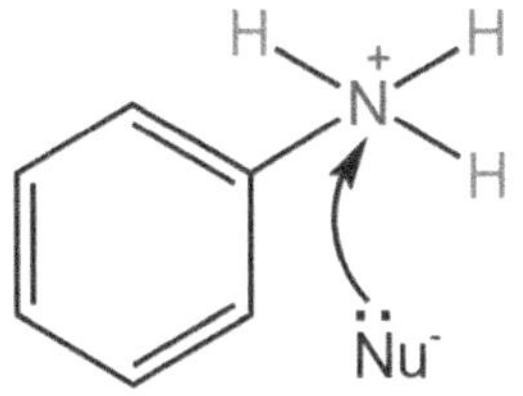

(d)
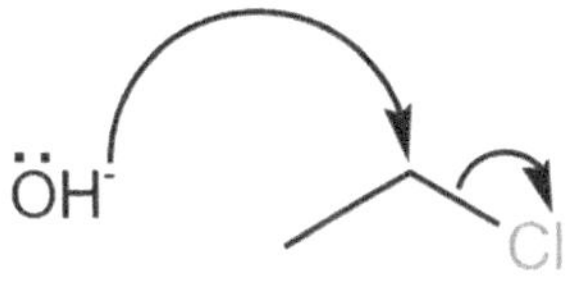

(e)
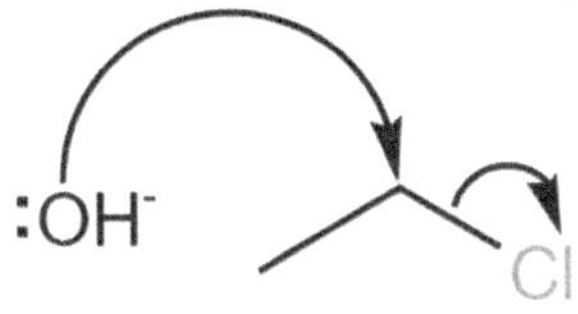

(f)
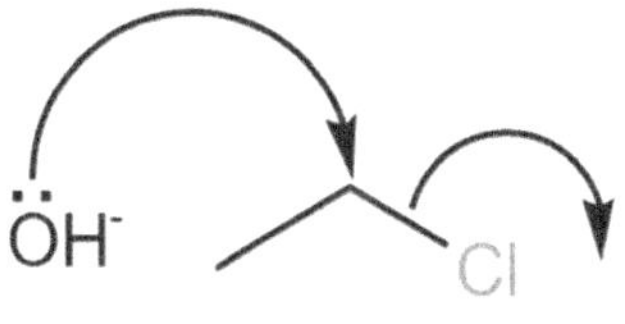

Solution:

- **(a)** Alkene π (Nu) → E^+ → full-headed arrow for electrophilic addition.

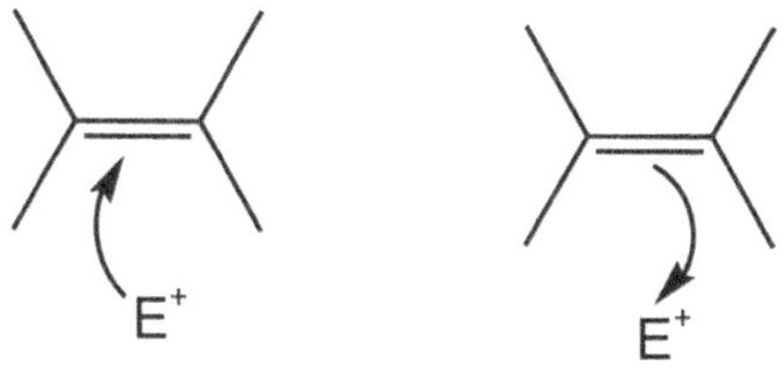

Wrong Correct

- **(b)** Nu^- abstracts H^+ from oxonium (deprotonation), not attack O.

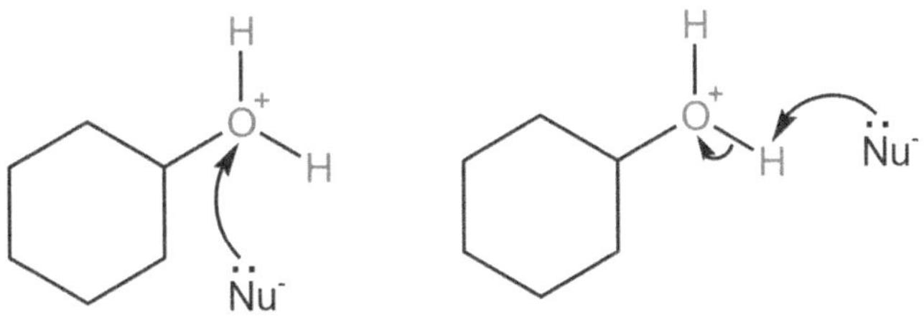

Wrong Correct

- **(c)** Nu^- deprotonates ammonium—no direct N^+ attack.

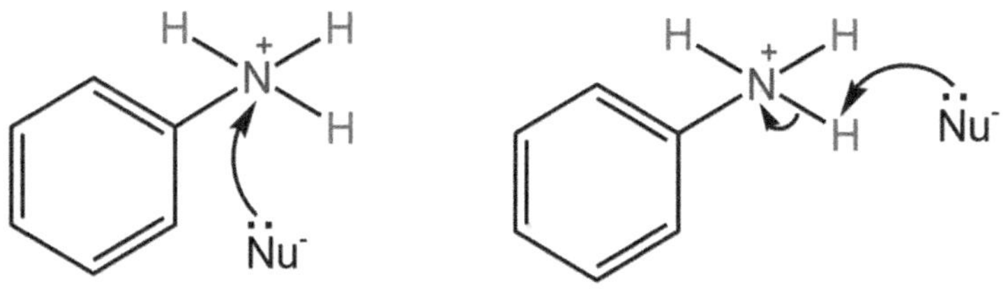

Wrong Correct

- **(d)** Arrows track **electrons**, not charges.

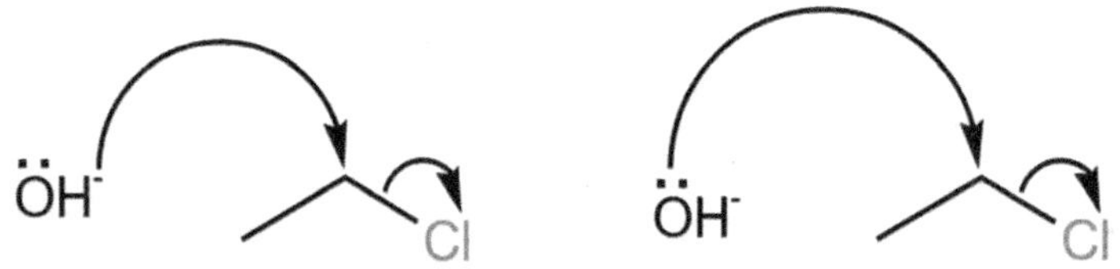

Wrong Correct

- **(e)** Arrows depict **electron flow**, not atom shifts.

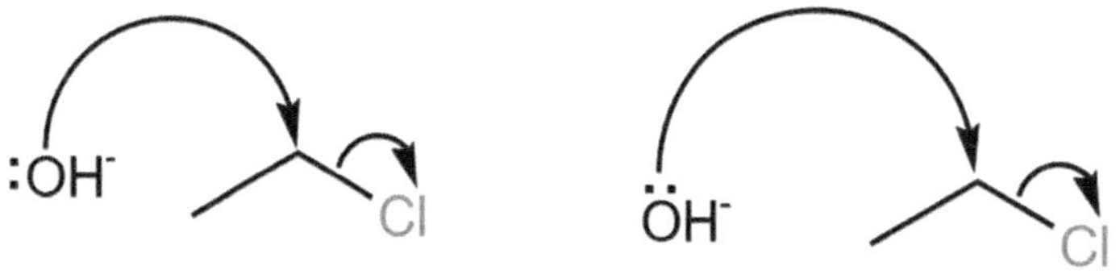

Wrong Correct

- **(f)** Heterolytic C–Cl cleavage: arrow from C–Cl bond → Cl, forming Cl^-.

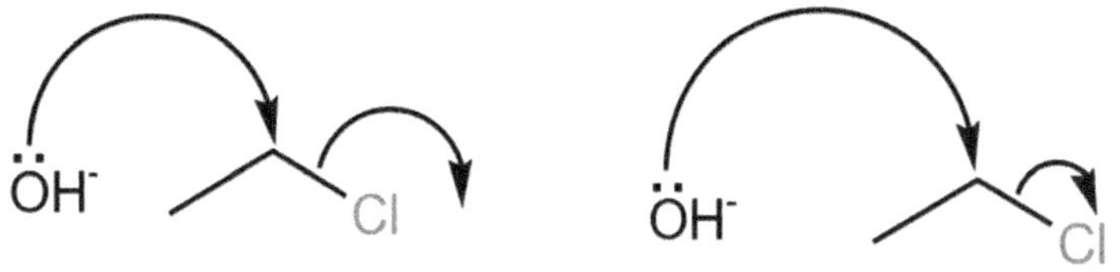

Wrong Correct

4.6 Putting It All Together

1. **Spot** Nu^- & E^+ (Ch. 2–3)

2. **Choose** the right arrow type

3. **Execute** core moves (attack, proton transfer, LG departure, rearrangement)

4. **Predict** the product

Master these steps and no mechanism—no matter how exotic—can evade your detective work.

Bet you didn't realize that by simply **spotting your players** and drawing a handful of arrows, you can crack any organic mechanism like a seasoned sleuth!

5 The Deduction Playbook — Mastering Structure Elucidation

"In the pursuit of knowledge, let your goal be more than just points—become a detective unraveling the mysteries of molecules. Each problem solved is a step toward mastering the art of organic chemistry."

— DT

5.1 A Question Type That Stumps the Best

Structure elucidation questions often appear in school exams—especially prelims—as long, multi-part problems. A typical 9 mark question might present **six compounds** with interlinked reactions. The good news? The **marking scheme is transparent**:

- **1 mark per correct structure**
- **0.5 mark per valid deduction**

That means if you identify six solid deductions and solve even one or two structures, you're already eyeing **–5 marks**—an excellent result for a high-level question.

5.2 Play the Game: Accuracy Meets Time Strategy

In exams, **time is your constraint**. Use this simple guide:

> **Time allocation = 1.5 × marks**
> So, a 9-mark question = ~13.5 minutes.

But during practice? **Don't rush.** Your goal is mastery, not speed. Slow, deliberate practice builds familiarity and intuition. Over time, what once felt like guesswork becomes structured reasoning.

5.3 Elucidation Is a Puzzle — Not a Panic Button

Can't see the answer immediately? Don't panic. **Skip, return, rethink.**
Big chunks of text can be intimidating, but remember:

> **"A picture is worth a thousand words."**
> Turn wordy descriptions into **reaction maps or flowcharts** during revision. This transforms messy prose into clear chemical logic.

5.4 Exam vs. Practice Mindset

- **During exams:**
  - Go for points.
  - Prioritize deductions.
  - Finish the paper first, then come back to tough ones.
- **During practice:**
  - Go for clarity.
  - Check every atom and charge.
  - Don't settle for half-understood answers.

5.5 DT's Top 8 Tips for Elucidation Success

- ☑ **Convert prose into flowcharts** — get a bird's-eye view
- ☑ **Work backward from the product** — what could've led here?
- ☑ **Use skeletal formulas** — save time, reduce clutter
- ☑ **Count atoms** — C, H, O, X, N; never assume
- ☑ **Stay open-minded** — avoid locking into one idea too early
- ☑ **No answer peeking!** — train your brain, not your memory
- ☑ **Draw mechanisms** — especially for unfamiliar reactions
- ☑ **Practice consistently** — intuition is built, not born

5.6 Final Words

Every structure elucidation question is a mystery waiting to be solved—and you're the sleuth. You won't just be rewarded with marks, but with a **deeper mastery** of organic chemistry. So treat each problem like a training case. With time, you won't just survive these questions—you'll start to enjoy cracking them wide open.

Part 2

Foundations of Deduction

Core skills and strategies to build clarity, precision, and confidence.

6 The Reagent Files — Interrogating Functional Groups

"In the labyrinth of organic compounds, every atom whispers secrets; it is our duty, dear Watson, to decode their silent confessions with the precision of a master sleuth."

— **DT**

6.1 The First Clue: Reagents and Conditions

Every successful investigation begins with the right tools—and in organic chemistry, those tools are your **reagents and conditions**. Before you can tackle multi-step synthesis or structure elucidation questions with finesse, you must first **master single-step transformations**. Just as Sherlock Holmes never overlooks a fingerprint or footprint, you must be alert to the clues each functional group provides.

6.2 Think Like a Detective: Functional Groups on the Stand

To interrogate a molecule, start by identifying the "suspects"—your core **functional groups**:

- **Alkanes**
- **Alkenes**
- **Alcohols**
- **Halogenoalkanes**
- **Carbonyl compounds** (aldehydes & ketones)
- **Carboxylic acid derivatives** (acyl chlorides, esters, anhydrides)
- **Nitrogen compounds** (amines, nitriles, amides)

Now, ask the critical question:

> "How can I convert *this* functional group into *that* one?"

For instance:

- Can I turn an **alkene into an alcohol**? Yes—via hydration.
- Can I go from **alkane to alcohol** in one step? Not directly.

Knowing **what's possible**—and what isn't—is just as important as knowing the how.

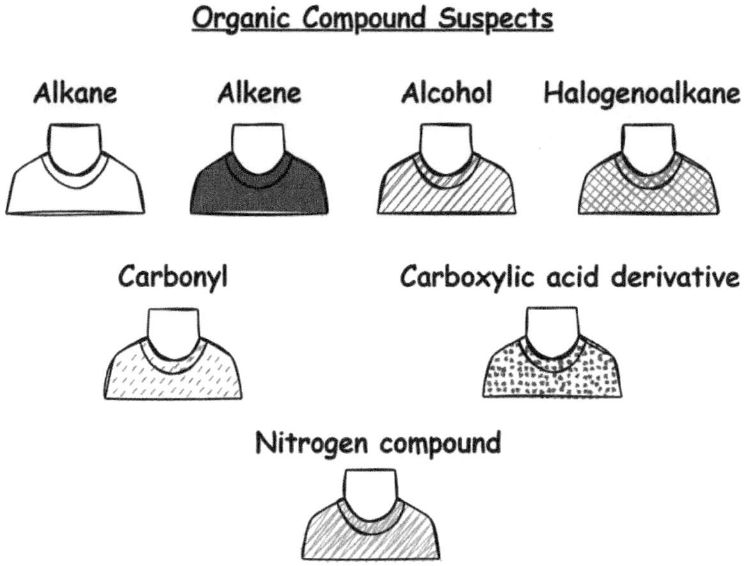

6.3 The Detective's Mindset: Mapping Connections

Students often rely on **mind maps** to remember transformations, but remember:

> A mind map is only powerful when drawn from memory—not copied.

Train yourself to mentally **link functional groups** through valid one-step conversions. Try this drill:

- Start with **alkanes**.
  Ask: Can I get to **alkenes, halogenoalkanes, alcohols**, etc.?

- Now work in reverse.
  From a **nitrile**, can I reach an **amine, carboxylic acid**, or **carbonyl**?

By moving both **forward and backward**, you reinforce mental "bridges" that help you recall pathways with speed and accuracy. These connections are the **dendrites** of your organic chemistry brain—strengthened through active recall.

6.4 Functional Groups Are Not All Equal

Not all functional groups behave independently—they come with subtypes and unique behavior:

- **Carbonyl compounds** include **aldehydes** and **ketones**, with different reactivity.
- **Carboxylic acid derivatives** form a family: **acyl chlorides, esters, amides**, etc.
- **Nitrogen compounds** split into **amines (1°, 2°, 3°)**, **nitriles**, and **amides**, each with specific reactivities and preferred conditions.

Like witnesses in a case, you must know their **individual stories**, but also how they're **connected**.

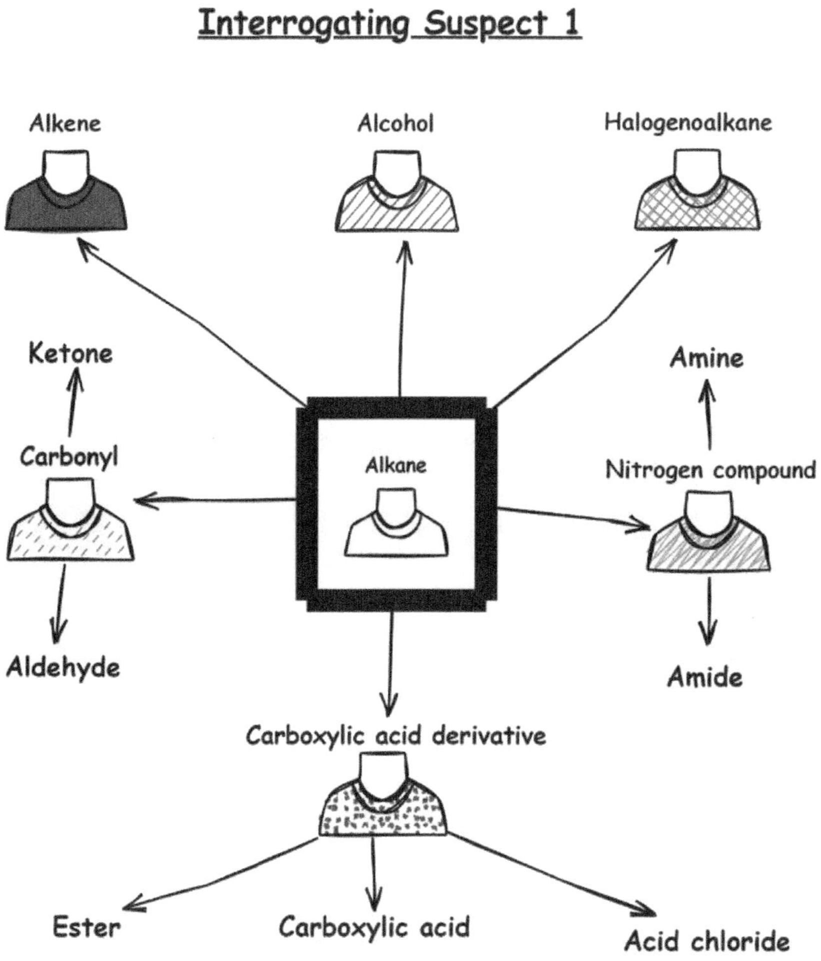

6.5 Why This Matters

Reagents and conditions are your **foundational knowledge**. Without it:

- Reaction mechanisms make no sense.
- Synthesis questions fall apart.
- Elucidation becomes guesswork.

With it, however, you gain **confidence and clarity**—the power to predict, transform, and solve.

6.6 Final Words from the Investigation Room

Organic chemistry isn't about brute-force memorisation—it's about recognising **patterns** and applying **logic**. Start by learning what each reagent does, which conditions it requires, and how functional groups respond under pressure.

> Just like a seasoned detective, **you don't just memorize suspects—you understand their behavior.**

So keep revisiting your reagent files. The more you interrogate them, the more they'll reveal.

---

7 The Chemistry of Clarity

"As one navigates the labyrinth of organic compounds, clarity emerges as both compass and map, illuminating the path where chaos once reigned."

— **DT**

To keep things crystal clear (and your sanity intact), I've split the flashcard-style guides into **aliphatic** and **aromatic** reactions. Trust me — lumping them together would be like mixing spaghetti with sushi. Both delicious, but together? Chaos.

These flashcards are your personal crash course. Go through them a few times and you'll start recognising the reagents and conditions the way a pianist recognises chords — effortlessly, instinctively, and with flair. That's the point: to get these reactions into your muscle memory so structure elucidation becomes second nature.

Now, let's address the elephant in the flask: the reagents and conditions listed aren't the only ones that exist. Organic chemistry is a generous subject — it offers multiple routes to the same destination. But if I packed them all in, the tables would be so cramped, you'd need a magnifying glass and a stress ball.

So I've chosen the clearest, cleanest, most exam-friendly options — the ones that get you marks without the mess. Once you've mastered these, you'll be well-equipped to handle whatever twists your paper throws at you.

Functional-Group Key

#	Functional Group	Typical examples
1	Alkane	CH_3-CH_3
2	Alkene	CH_2=CH_2
3	Arene (benzene / alkylbenzene)	C_6H_6, CH_3-C_6H_5
4	Alcohol / Phenol	CH_3CH_2OH, C_6H_5OH
5	Halogenoalkane / Aryl halide	CH_3CH_2Br, C_6H_5Cl
6	Carbonyl (Aldehyde/Ketone)	CH_3CHO, CH_3COCH_3
7	Carboxylic Acid	CH_3COOH
8	Acid Chloride	CH_3COCl
9	Ester	CH_3COOCH_3
10	Amide	CH_3CONH_2
11	Amine	$CH_3CH_2NH_2$, $C_6H_5NH_2$
12	Nitrile	CH_3CN

Aliphatic One-Step Conversion Guide

From # → To #	Minimal reagent / condition	From # → To #	Minimal reagent / condition
1 → 2	Al_2O_3, 500 °C (cracking)	2 → 1	H_2, Ni, 150 °C
1 → 5	Cl_2, hv	5 → 2	KOH (ethanolic), heat
2 → 5	HX, r.t.	5 → 4	NaOH (aq), reflux
2 → 4	H_2O/H^+, 300 °C, 70 atm	4 → 2	conc H_2SO_4, heat
4 → 5	$SOCl_2$ (or PCl_5)	4 → 6	$K_2Cr_2O_7/H^+$, distil (1°) / reflux (2°)
6 → 4	$NaBH_4$, r.t.	6 → 7	$K_2Cr_2O_7/H^+$, reflux (aldehyde)
6 → 11	HCN (add) → $LiAlH_4$ (reduce)	7 → 8	$SOCl_2$, r.t.
7 → 9	ROH, conc H_2SO_4, reflux	7 → 4	$LiAlH_4$ then H_2O
8 → 9	ROH, r.t.	8 → 10	NH_3, r.t.
9 → 7	H^+/H_2O, heat	10 → 11	$LiAlH_4$ then H_2O
10 → 7	H^+/H_2O, heat	11 → 10	RCOCl, r.t.
12 → 11	$LiAlH_4$ then H_2O	12 → 7	H^+/H_2O, heat
5 → 11	NH_3, ethanol, pressure	5 → 12	KCN, ethanol

Aliphatic Drill Grid — Fill the Reagents

From # → To #	Minimal reagent / condition	From # → To #	Minimal reagent / condition
1 → 2		2 → 1	
1 → 5		5 → 2	
2 → 5		5 → 4	
2 → 4		4 → 2	
4 → 5		4 → 6	
6 → 4		6 → 7	
6 → 11		7 → 8	
7 → 9		7 → 4	
8 → 9		8 → 10	
9 → 7		10 → 11	
10 → 7		11 → 10	
12 → 11		12 → 7	
5 → 11		5 → 12	

Aromatic One-Step Conversion Guide

From / Intermediate	To	Minimal reagent / condition	Notes
3 (Benzene)	Alkylbenzene (still 3)	RCl, $AlCl_3$	Friedel-Crafts alkylation (electrophilic subst)
3	5 (Chlorobenzene)	Cl_2, $AlCl_3$	Electrophilic chlorination (electrophilic subst)
3	Nitrobenzene	conc HNO_3/H_2SO_4	Nitration (electrophilic subst)
Nitrobenzene	11 (Phenylamine)	Sn/HCl (or H_2/Ni)	Reduction
Alkylbenzene (3)	7 (Benzoic acid)	$KMnO_4$, heat	Side-chain oxidation
4 (Phenol)	9 (Phenyl ester)	+ Acyl chloride, r.t.	Acylation of phenol (nucleophilic acyl subst)
11 (Phenylamine)	10 (Amide)	+ Acyl chloride	Nucleophilic acyl subst

Aromatic Drill Grid — Fill the Reagents

From / Intermediate	To	Minimal reagent / condition	Notes
3 (Benzene)	Alkylbenzene (still 3)		
3	5 (Chlorobenzene)		
3	Nitrobenzene		
Nitrobenzene	11 (Phenylamine)		
Alkylbenzene (3)	7 (Benzoic acid)		
4 (Phenol)	9 (Phenyl ester)		
11 (Phenylamine)	10 (Amide)		

8 Skeleton Key: Unlocking Efficient Isomer Drawing

"The true art of structure elucidation lies not merely in observation but in contemplation of the skeletal form, where every line and vertex whispers tales of connectivity and character yet to be discovered."

— **DT**

8.1 Why Skeletal Formulas Matter

In structure elucidation questions, you're often asked to deduce the identity of unknown compounds using chemical tests and molecular formulae. While **displayed formulas** are commonly taught in school, **skeletal formulas** are an **essential tool**—especially when identifying **isomers**.

Sadly, skeletal formulas are often skimmed over in class. But with a little practice, they can **dramatically speed up your thought process** and reduce clutter when dealing with complex molecules.

8.2 Let's Investigate an Example

You're told:

Compound **A** has the molecular formula $C_4H_{10}O$. It reacts with sodium metal, producing a gas that extinguishes a lighted splint with a 'pop' sound. When acidified $KMnO_4$ is added to Compound **A**, it forms a compound **B** that yields an orange precipitate with 2,4-DNPH. Deduce the structure of Compound **A**.

Let's work like detectives:

Observation	Deduction
1. Compound **A** reacts with Na to produce a gas.	Compound **A** is an alcohol and the gas produced is H_2.
2. Compound **A** reacts with acidified $KMnO_4$	Compound **A** is either a primary or a secondary alcohol.
3. Compound **B** reacts with 2,4-DNPH	Compound **B** is either an aldehyde or ketone. Since $KMnO_4$ was used to produce compound **B**, **B** cannot be an aldehyde. **B** is a ketone.

Since compound **B** is a ketone, compound **A** must be a secondary alcohol.

8.3 Why Systematic Drawing Is Essential

Knowing **A** is a secondary alcohol is not the end—you still need to **draw its correct structure**, and perhaps the structure of **B** too.

Most students now **guess** and draw randomly. But without a strategy, this leads to wasted time and missed isomers.

Instead, here's a structured approach:

8.4 DT's 3-Step Method for Drawing Isomers

Step 1: Draw the Carbon Skeleton

Start by placing the correct number of carbon atoms (based on the molecular formula). For **$C_4H_{10}O$**, try both straight and branched chains.

n = 4:

n = 3:

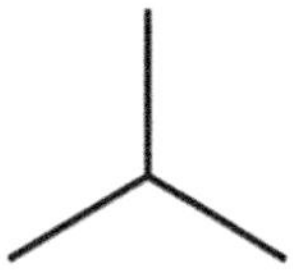

Step 2: Add the Functional Group

Once the backbone is built, add your **–OH**, **C=O**, **NH_2**, etc., where appropriate. Consider **primary**, **secondary**, and **tertiary** positioning.

n = 4:

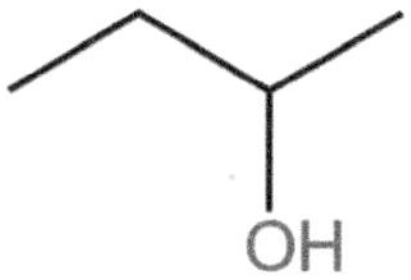

(secondary alcohol)

n = 3:

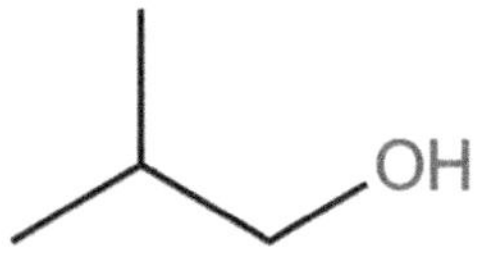

Step 3: Refine and Verify

Check for duplicates, valency errors, or unrealistic structures. Make sure the molecule aligns with the data and **chemical behaviour** described.

Compound A:

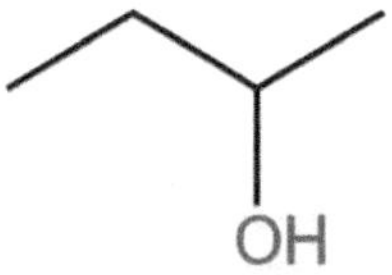

Compound B:

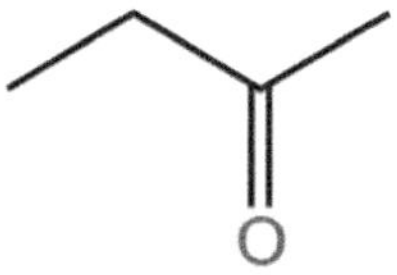

8.5 The Real Challenge: Bigger Molecules

As molecules grow in size, the number of possible isomers increases dramatically. Without a **systematic plan**, the risk of missing or repeating structures becomes almost inevitable.

That's why this skill—drawing isomers accurately using **skeletal formulas** and a logical method—is so powerful. It helps you move **beyond trial and error** and into confident, controlled reasoning.

8.6 Final Thought

Structure elucidation isn't just about tests and functional groups. It's about **pattern recognition** and **strategy**—and your drawing method plays a key role in this. Skeletal formulas are your **shortcuts**, and this 3-step method is your **compass**.

> Use the Skeleton Key. Never guess. Always deduce.

---

9 Decode with DOU — Your Unsaturation Toolkit

"To navigate the intricate web of organic compounds is to wield unsaturation as your most trusted ally; it transforms chaos into order, revealing truths that are often obscured by mere appearances."

—DT

9.1 The Hidden Clue in Every Formula

In every great detective story, there's a small, easily overlooked clue that reveals the truth — if you know how to spot it. In organic chemistry, that clue is the **Degree of Unsaturation (DOU)**, also known as the **Index of Hydrogen Deficiency (IHD)**.

This calculation reveals how many **rings**, **double bonds**, or **triple bonds** a molecule contains — just from its molecular formula. It's a fast, powerful way to filter through possible structures without counting atoms one by one.

Example: C_8H_8O

- DOU = **5**
- Suggests a **benzene ring (4 DOU)** and **one additional unsaturation**, likely a **C=O**.
- Instantly narrows down functional group possibilities.

9.2 Why DOU is So Useful

When used alongside spectroscopic data (like **IR**, **NMR**, and **MS**) and chemical tests, the DOU becomes your **first step in structure elucidation**. It provides direction, saves time, and improves accuracy.

Rather than blindly guessing, you can **think like a detective** and work from known facts.

9.3 How to Calculate the Degree of Unsaturation

9.3.1 The General Formula

For molecules containing carbon (C), hydrogen (H), nitrogen (N), oxygen (O), and halogens (X), the full formula for calculating the **Degree of Unsaturation (DOU)** is:

DOU = (2C + 2 + N - H - X) ÷ 2

Where:

- **C** = number of carbon atoms
- **H** = number of hydrogen atoms
- **N** = number of nitrogen atoms
- **X** = number of halogen atoms (F, Cl, Br, I)

This formula is accurate and works in all cases — but let's be honest: **it's not the easiest to remember**, especially under exam stress. The brackets, the halogens, the nitrogens... it can feel overwhelming when time is tight.

That's why we recommend a **simplified shortcut** for A Level exams — one that works just as well and is much easier to apply quickly.

Let's look at this shortcut next.

9.3.2 A Simpler Shortcut (Modified Hydrogen Count Method)

For hydrocarbons:

DOU = C - H/2 + 1

Adjust for other atoms:

- **Halogen (X):** Add 1 H for each halogen
- **Oxygen (O):** Ignore
- **Nitrogen (N):** Subtract 1 H for each N

Then plug into the shortcut formula using the modified hydrogen count.

9.4 Practice: Worked Examples

Example 1: C_2H_5Cl

- Replace Cl with H → C_2H_6
- DOU = 2 - (6 ÷ 2) + 1 = **0**

Example 2: C_3H_6O

- Ignore O → C_3H_6
- DOU = 3 - (6 ÷ 2) + 1 = **1**

Example 3: C_2H_3N

- Subtract 1 H for N → C_2H_2
- DOU = 2 - (2 ÷ 2) + 1 = **2**

9.5 Interpreting DOU: What It Tells You

Each structural feature contributes to the total DOU:

Structure Type	**DOU Contribution**
Ring	1
Double bond (C=C, C=O, C=N)	1
Triple bond (C≡C, C≡N)	2
Benzene ring	4

9.6 DOU Summary Table

DOU Value	Possible Interpretation
0	Fully saturated (no rings or multiple bonds)
1	One ring or one double bond
2	One triple bond, or two rings/double bonds
3	Multiple combinations of rings and double bonds
4	Likely a benzene ring
>4	Benzene ring plus additional unsaturations

⚠ **Note:** A DOU of 4 suggests, but does not confirm, aromaticity.

9.7 Aromatic ≠ DOU of 4

A **common mistake** is assuming:

"If DOU = 4, the molecule is aromatic." ✖

This is not always true. **Furan** has a DOU of 3 but is aromatic. Aromaticity depends on **π-electron count**, following **Hückel's Rule (4n + 2 π-electrons)**, not DOU.

9.8 Estimating DOU from Displayed Structures

Count:

- **1** for each ring or double bond
- **2** for each triple bond
- **4** for each benzene ring

Example Structures:

(a)

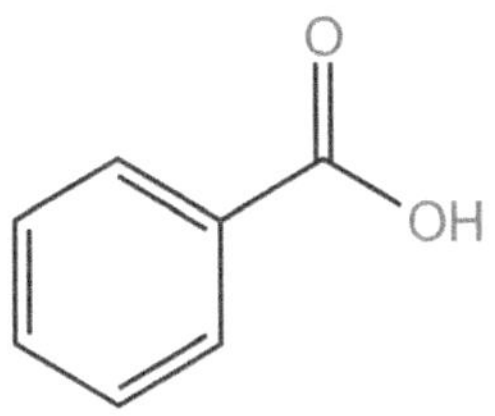

(b)

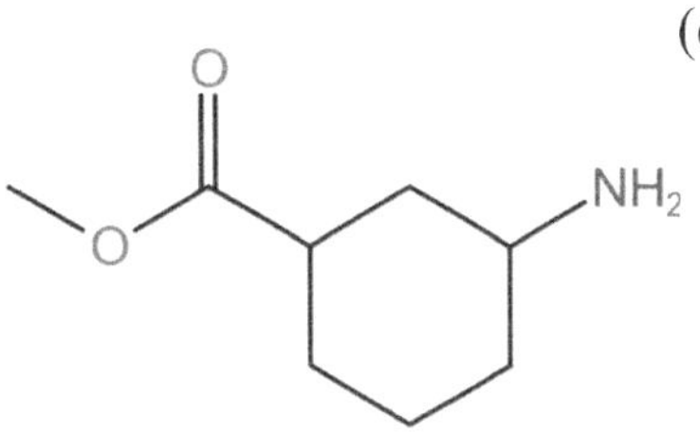

(c)

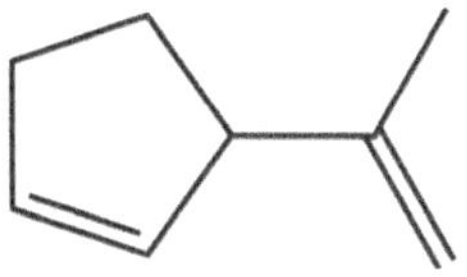

(d)

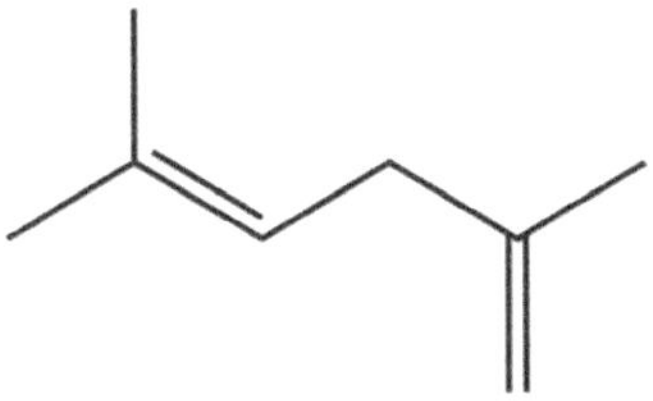

(e)

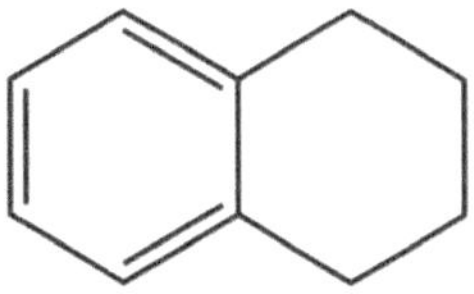

(f)

Structure	DOU Breakdown	Total DOU
(a)	Benzene ring (4) + C=O (1)	**5**
(b)	One ring + one double bond	**2**
(c)	One ring + two double bonds	**3**
(d)	Two double bonds	**2**
(e)	Benzene ring (4) + ring (1)	**5**
(f)	Ring + two double bonds	**3**

9.9 Case Study: $C_4H_6OCl_2$

Step 1: Calculate DOU

Using the general formula:
DOU = (2×4 + 2 - 6 - 2) ÷ 2 = **1**

Possible explanations:

- One ring
- One double bond (C=O or C=C)

Suppose you draw two possible structures:

- **Structure A:** DOU = **1** ✅
- **Structure B:** DOU = **2** ✕

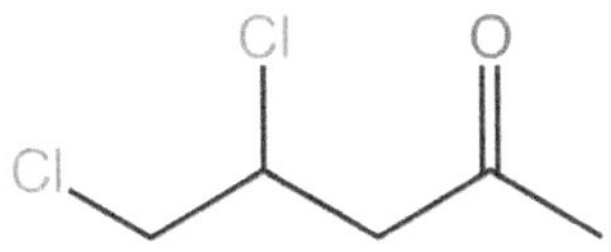

Structure A

Or

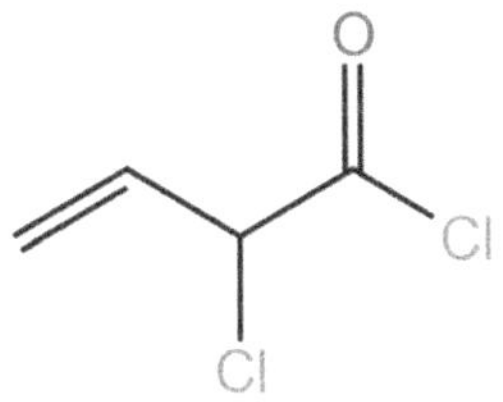

Structure B

Conclusion: Only Structure A matches the formula. No need to count hydrogens — DOU confirms it more efficiently.

9.10 Conclusion: DOU as Your First Step

The **Degree of Unsaturation (DOU)** is one of the most **effective tools** in your organic chemistry toolbox. It helps you:

- **Skip the guesswork**
- Work from **molecular formula → unsaturation → functional group clues**
- **Save time**, **avoid errors**, and draw **accurate structures** fast

DOU is the key that unlocks the hidden blueprint of a molecule — use it **wisely**, and you'll **reveal truths** that others overlook.

Part 3

The Quest for Perfection

Full-length deduction problems, increasing in complexity by element.

10 C, H, and O — A Systematic Start

"In the intricate dance of carbon, hydrogen, and oxygen, clarity emerges from chaos. Every molecular puzzle holds a story waiting to be unveiled." — **DT**

In **Chapter 9**, we saw how the **Degree of Unsaturation (DOU)** can act as a powerful shortcut to unraveling unknown molecular structures. Now, we begin applying that insight to full organic questions — the kind where you're given **a series of compounds**, some test results, and the instruction to **work out their structures**.

But here's the twist — this book doesn't follow the **BANE** method you might see in school prelim answers, with mechanical tables like:

Observation	Deduction
• **A** undergoes elimination with conc. H_2SO_4 to give **B**.	• **A** is an alcohol. • Since **A** only has 2 C atoms, it must be ethanol.
• **B** undergoes electrophilic addition with Br_2 in CCl_4.	• **B** is an alkene. • **B** is ethene.

Instead, I'm guiding you through my **actual thought process** — how I read a question, what I focus on, and how I use clues (sometimes subtle ones) to logically link one compound to the next. You'll follow a **prose-style walkthrough** of how I would approach the problem, almost like sitting beside me while we solve it together.

A Layered Approach to Complexity

Organic chemistry can seem chaotic at first glance — so many reactions, reagents, and mechanisms. But chaos isn't random. In fact, as with **chaos theory**, there's hidden order behind the complexity. Our aim is to uncover that order, one logical step at a time.

To make this journey more manageable, the next three chapters will **gradually increase in complexity**:

- **Chapter 10**: Compounds involving only **carbon, hydrogen, and oxygen (C, H, O)**
- **Chapter 11**: Introducing **halogens (X)** — for compounds like haloalkanes and acid derivatives
- **Chapter 12**: Adding **nitrogen (N)** — to include amines, amides, nitriles, etc.

This **incremental approach** lets you build confidence with simpler systems before tackling multi-functional compounds. Importantly, it also lets you learn how to spot **subtle differences** in question phrasing. Some structure elucidation questions may look almost identical — but a single phrase like:

- *"Gives a positive Fehling's test"*
- vs.
- *"Does not react with Fehling's solution"*

...can lead to **completely different structures** (aldehyde vs. ketone). We'll train your eye to pick up on these nuances.

How This Book Works

Each question is approached as a **puzzle**, not a checklist. You'll see me explain — in words — how I interpret the clues, what I prioritise, how I sketch possible structures in my mind, and how I validate them using DOU and reactivity patterns.

And at the end of each question?

You'll get a **flow chart** — a visual summary that links all the compounds mentioned in the question. These flow charts are designed for **fast, intuitive understanding**:

- They don't repeat the full reaction conditions (which are already in the question).
- Instead, they show the **interconnection between compounds** — how each one transforms into the next in a logical sequence.

This gives the book **two powerful functions**:

1. **Full Problem Solving** — Use pen and paper to work through the structure from scratch, checking your answers with the flow chart at the end.
2. **Quick Reference or Revision on the Go** — While travelling or revising, just study the flow chart and mentally work out the reagents or conditions needed to go from one compound to the next.

In short, this isn't just a book of questions and answers. It's a **thinking guide** — a way to train your brain to approach organic chemistry like a detective, spotting clues, linking ideas, and seeing the big picture.

Ready to begin?

Let's now explore the world of organic problems involving just **C, H, and O**. From **alcohols** to **carbonyls**, and **carboxylic acids** to **esters**, we'll see how a solid understanding of functional groups and DOU can illuminate even the trickiest of puzzles.

Time to let the logic unfold.

Question 1:

Compound **A** has the molecular formula C_4H_8. When treated with cold concentrated sulfuric acid, it yields compound **B**, which has the formula $C_4H_{10}O$. Compound **B** subsequently undergoes oxidation when reacted with acidified sodium dichromate for prolonged period to produce compound **C**, C_4H_8O. Additionally, compound **A** is capable of decolorizing liquid bromine. Furthermore, upon reaction with hot acidified potassium permanganate(VII), compound **A** generates compound **D**, $C_3H_6O_2$.

Propose structures for compounds **A**, **B**, **C** and **D**, and discuss the observations made during these reactions.

Question 2:

Compound **A** has a molecular formula of C_5H_{10} and is known to decolorize aqueous bromine. When compound **A** reacts with acidified potassium permanganate(VII), it yields compound **B**, which has the formula C_4H_8O. Furthermore, compound **B** reacts with 2,4-dinitrophenylhydrazine, resulting in the formation of an orange precipitate.

Propose structures for compounds **A**, **B**, and discuss the observations made during these reactions.

Question 3:

Compound **A** has a molecular formula of C_5H_{10} and is capable of decolorizing liquid bromine. Compound **A** also reacts with potassium permanganate(VII) to produce compound **B**, which has the formula $C_4H_8O_2$, while releasing a gas that forms a white precipitate with calcium hydroxide. It is noteworthy that compound **B** does not yield an orange precipitate when treated with Brady's reagent. Additionally, compound **B** can be reduced by lithium aluminium hydride ($LiAlH_4$) to form compound **C**, which has the formula $C_4H_{10}O$. Upon treatment with hot concentrated sulfuric acid, compound **C** yields compound **D** (C_4H_8). When subjected to oxidation, compound **D** produces compound **E** ($C_3H_6O_2$), which generates effervescence upon exposure to sodium hydrogen carbonate.

Propose structures for compounds **A** through **E** and elucidate the observations described in the reactions.

Question 4:

Compound **A** has a molecular formula of C_5H_{10} and can decolorize aqueous bromine. Upon treatment with potassium permanganate(VII), it produces compound **B**, which has the formula $C_4H_8O_2$. There is no observable reaction between compound **B** and 2,4-dinitrophenylhydrazine. When lithium aluminium hydride is added to compound **B**, it yields compound **C**, with the formula $C_4H_{10}O$. The addition of excess hot concentrated sulfuric acid to compound **C** results in the formation of compound **D,** C_4H_8.

When subjected to strong oxidation, compound **D** generates compound **E** along with effervescence. Additionally, compound **E** forms an orange precipitate when treated with Brady's reagent.

Suggest structures for compounds **A** through **E** and explain the observations described above.

Question 5:

Compound **A** has a molecular formula of C_5H_{10}, which is capable of decolorizing bromine water. When subjected to acidified potassium permanganate(VII), it produces two organic compounds: compound **B** with the formula $C_2H_4O_2$ and compound **C** with the formula $C_3H_6O_2$. Notably, both compounds **B** and **C** do not exhibit any observable change when treated with 2,4-dinitrophenylhydrazine.

Suggest structures for compounds **A**, **B**, and **C**, and provide an explanation for the observations described above.

Question 6:

Compound **A** has the molecular formula C_5H_{10}, which decolorizes bromine water upon treatment. When subjected to acidified potassium permanganate(VII), it yields two organic compounds: compound **B** with the formula C_3H_6O and compound **C** with the formula $C_2H_4O_2$. Only compound **B** forms an orange precipitate when treated with Brady's reagent.

Suggest structures for compounds **A**, **B**, and **C**, and provide a detailed explanation for the observations described above.

Question 7:

Compound **A** has the molecular formula C_3H_8O. It reacts with hot concentrated phosphoric acid to yield compound **B**, which has the formula C_3H_6 and rapidly decolorizes a solution of bromine in tetrachloromethane. When treated with ethanoyl chloride, compound **A** produces compound **C**, with the molecular formula $C_5H_{10}O_2$. Additionally, when compound **A** is oxidized using potassium dichromate in dilute sulfuric acid, it forms compound **D**, C_3H_6O, which gives an orange precipitate with 2,4-dinitrophenylhydrazine but does not react with an ammoniacal solution of silver nitrate.

Deduce the structures of compounds **A**, **B**, **C**, and **D**, and explain carefully what each piece of evidence tells you about these compounds.

Question 8:

Compound **A** has the molecular formula C_3H_8O. It reacts with hot concentrated phosphoric acid to yield compound **B**, which has the formula C_3H_6 and rapidly decolorizes a solution of bromine in tetrachloromethane. When treated with ethanoyl chloride, compound **A** produces compound **C**, with the molecular formula $C_5H_{10}O_2$. Additionally, when compound **A** is oxidized using potassium dichromate in dilute sulfuric acid, it forms compound **D** ($C_3H_6O_2$), which gives effervescence with potassium hydrogen carbonate.

Deduce the structures of compounds **A**, **B**, **C**, and **D**, and explain carefully what each piece of evidence tells you about these compounds.

Question 9:

Compound **A** ($C_4H_{10}O$) does not react with potassium dichromate(VI). When heated with concentrated sulfuric acid, it forms compound **B** (C_4H_8), which causes the decolorization of aqueous bromine.

Compound **B** reacts with acidified potassium permanganate(VII) to produce compound **C** (C_3H_6O), accompanied by the evolution of a gas that forms a white precipitate when bubbled through calcium hydroxide solution.

Deduce the structures of compounds **A**, **B**, and **C**, and explain carefully what each piece of evidence suggests about their identities.

Question 10:

When treated with sodium dichromate(VI), compound **A** ($C_4H_{10}O$) forms a product that does not react with 2,4-dinitrophenylhydrazine but produces effervescence upon the addition of potassium carbonate.

When heated with concentrated sulfuric acid, compound **A** forms compound **B** (C_4H_8), which decolorizes liquid bromine.

Upon reaction with hot acidified potassium permanganate(VII), compound **B** is converted into compound **C** (C_3H_6O), with effervescence observed, and the resulting gas produces a white precipitate when bubbled through calcium hydroxide solution.

Deduce the structures of compounds **A**, **B**, and **C**, and explain carefully what each piece of evidence suggests about their identities.

Question 11:

Compound **A** ($C_4H_{10}O$) reacts with potassium dichromate(VI) to form a product that does not produce an orange precipitate when treated with 2,4-dinitrophenylhydrazine.

When heated with concentrated sulfuric acid, compound **A** forms compound **B** (C_4H_8) only, which decolorizes liquid bromine.

Upon reaction with hot acidified potassium permanganate(VII), compound **B** is converted into compound **C** ($C_3H_6O_2$), with the evolution of a gas.

Deduce the structures of compounds **A**, **B**, and **C**, and explain carefully what each piece of evidence suggests about their identities.

Question 12:

Upon heating with potassium dichromate(VI) for a prolonged period, compound **A** is oxidized to compound **E**, which gives an orange precipitate with 2,4-dinitrophenylhydrazine.

When heated with concentrated sulfuric acid, compound **A** produces three isomeric alkenes, **B**, **C**, and **D**, all having the molecular formula C_4H_8.

Only compounds **C** and **D**, when treated with hot acidified potassium permanganate(VII), yield the same oxidation product, compound **F** ($C_2H_4O_2$); compound **B** gives organic compound **G**.

Deduce the structures of compounds **A**, **B**, **C**, **D**, **E**, **F**, and **G**, and explain carefully what each piece of evidence suggests about their identities.

Question 13:

Compound **A** ($C_5H_{12}O$) forms a yellow precipitate when treated with alkaline aqueous iodine. When oxidized with potassium dichromate(VI), it forms compound **E**.

When heated with concentrated sulfuric acid, compound **A** forms a pair of geometric isomers, **B** and **C**, and an additional isomer, **D**, all of which decolorize reddish-brown liquid bromine.

Upon reaction with hot acidified potassium permanganate(VII),

- Geometric isomers **B** and **C** yield the same two oxidation products, **F** ($C_3H_6O_2$) and **G** ($C_2H_4O_2$).
- Isomer **D** forms a single organic product, **H** ($C_4H_8O_2$), with the evolution of a gas.

Deduce the structures of compounds **A**, **B**, **C**, **D**, **E**, **F**, **G**, and **H**, and explain carefully what each piece of evidence suggests about their identities.

Question 14:

Compound **A** ($C_5H_{12}O$) reacts with potassium dichromate(VI) under reflux to form compound **F** ($C_5H_{10}O$), which produces an orange precipitate when treated with 2,4-dinitrophenylhydrazine.

When heated with concentrated sulfuric acid, compound **A** produces a pair of stereoisomers, compounds **B** and **C**, both of which decolorize liquid bromine.

Upon oxidation, both compound **B** and compound **C** yield the same two organic products, compound **D** ($C_2H_4O_2$) and compound **E** ($C_3H_6O_2$).

Deduce the structures of compounds **A**, **B**, **C**, **D**, **E**, and **F**, and explain carefully what each piece of evidence suggests about their identities.

Question 15:

When treated with potassium dichromate(VI), compound **A** ($C_5H_{12}O$) forms a product that produces an orange precipitate with 2,4-dinitrophenylhydrazine.

Upon reaction with concentrated sulfuric acid, compound **A** forms two isomers, **B** and **C**, both having the molecular formula C_5H_{10}. Both compounds decolorize liquid bromine.

When heated with hot acidified potassium permanganate(VII):

- Compound **B** forms compound **D** ($C_4H_8O_2$), with the evolution of a gas.
- Compound **C** forms two products: compound **E** ($C_2H_4O_2$) and compound **F** (C_3H_6O).

Deduce the structures of compounds **A**, **B**, **C**, **D**, **E**, and **F**, and explain carefully what each piece of evidence suggests about their identities.

Question 16:

Compound **A** ($C_5H_{12}O$) reacts with potassium dichromate(VI) to form compound **D** ($C_5H_{10}O_2$).

When heated with concentrated sulfuric acid, compound **A** forms compound **B** (C_5H_{10}), which decolorizes liquid bromine.

Upon reaction with acidified potassium permanganate(VII), compound **B** is converted into compound **C** ($C_4H_8O_2$) with the evolution of a gas.

When treated with lithium aluminium hydride, compound **C** forms compound **E** ($C_4H_{10}O$).

Upon heating with hot concentrated sulfuric acid, compound **E** forms compound **F** (C_4H_8).

When treated with hot acidified potassium permanganate(VII), compound **F** is oxidized to compound **G** ($C_3H_6O_2$).

Deduce the structures of compounds **A**, **B**, **C**, **D**, **E**, **F**, and **G**, and explain carefully what each piece of evidence suggests about their identities.

Question 17:

Compound **A** ($C_5H_{12}O$) is converted to compound **B** (C_5H_{10}) when treated with concentrated sulfuric acid. Compound **B** decolorizes liquid bromine.

When treated with potassium permanganate(VII), compound **B** produces a single organic product, **C** (C_4H_8O).

Deduce the structures of compounds **A**, **B**, and **C**, and explain carefully what each piece of evidence suggests about their identities.

Question 18:

Compound **A** ($C_5H_{12}O$) does not react with hot concentrated sulfuric acid.

When treated with sodium metal in dry ether, effervescence is observed, and the gas produced extinguishes a lighted splint with a 'pop' sound.

Deduce the structure of compound **A**, and explain carefully what each piece of evidence suggests about its identity.

Question 19:

Compound **A** ($C_5H_{12}O$) does not react with potassium dichromate(VI).

When heated with hot concentrated sulfuric acid, compound **A** forms two different compounds, **B** and **C**, both having the molecular formula C_5H_{10}.

Upon treatment with hot acidified potassium permanganate(VII):

- Compound **B** forms compound **D** ($C_2H_4O_2$) and compound **E** (C_3H_6O).
- Compound **C** forms only one organic compound, **F** (C_4H_8O).

When treated with 2,4-dinitrophenylhydrazine, only compounds **E** and **F** give a positive test, while compound **D** does not.

Deduce the structures of compounds **A**, **B**, **C**, **D**, **E**, and **F**, and explain carefully what each piece of evidence suggests about their identities.

Question 20:

Compound **A** ($C_5H_{12}O$) undergoes treatment with concentrated sulfuric acid, yielding compound **B** (C_5H_{10}), which decolorizes aqueous bromine.

When hot acidified potassium permanganate(VII) is added to compound **B**, it forms compound **C** ($C_4H_8O_2$), which does not form an orange precipitate with 2,4-dinitrophenylhydrazine.

When compound **A** is treated with potassium dichromate(VI), it forms compound **D** ($C_5H_{10}O_2$).

Upon treatment with lithium aluminium hydride, compound **C** is reduced to compound **E** ($C_4H_{10}O$).

When concentrated sulfuric acid is added to compound **E**, it forms compound **F** (C_4H_8).

When treated with hot acidified potassium permanganate (VII), compound **F** produces a single organic compound **G** (C_3H_6O).

Deduce the structures of compounds **A**, **B**, **C**, **D**, **E**, **F**, and **G**, and explain carefully what each piece of evidence suggests about their identities.

Question 21:

Compound **A** ($C_4H_8O_2$) is heated with sodium hydroxide, producing compound **B** (CH_4O) and an organic salt.

Upon acidic hydrolysis, the organic salt is converted into compound **C**.

When compound **C** is reduced, it forms compound **D**.

Upon heating with concentrated sulfuric acid, compound **D** forms compound **E**.

When compound **E** is treated with hot acidified potassium permanganate(VII), it is converted into compound **F** ($C_2H_4O_2$), which produces effervescence when reacted with sodium hydrogen carbonate.

Deduce the structures of compounds **A**, **B**, **C**, **D**, **E**, and **F**, and explain carefully what each piece of evidence suggests about their identities.

Question 22:

Compound **A** ($C_4H_8O_2$) is heated with sodium hydroxide, followed by acidification, to form compound **B** and compound **C**.

When hot acidified potassium permanganate(VII) is added to compound **B**, only effervescence is observed.

When hot acidified potassium permanganate(VII) is added to compound **C**, it is converted into compound **D**, which does not form an orange precipitate with 2,4-dinitrophenylhydrazine.

Additionally, compound **C** does not form a yellow precipitate when treated with alkaline aqueous iodine.

Deduce the structures of compounds **A**, **B**, **C**, and **D**, and explain carefully what each piece of evidence suggests about their identities.

Question 23:

Compound **A** ($C_4H_8O_2$) is heated with aqueous sodium hydroxide, forming compound **B** and compound **C**.

When hot acidified potassium permanganate(VII) is added to compound **B**, only effervescence is observed.

Compound **C** gives a yellow precipitate when treated with hot alkaline iodine.

Deduce the structures of compounds **A**, **B**, and **C**, and explain carefully what each piece of evidence suggests about their identities.

Question 24:

Compound **A** ($C_4H_8O_2$) is heated with aqueous sodium hydroxide, forming compound **B** and compound **C**.

Compound **C** gives a yellow precipitate when treated with hot alkaline iodine.

When compound **C** is treated with concentrated sulfuric acid, it forms an organic compound that produces only effervescence upon addition of hot acidified potassium permanganate(VII).

Deduce the structures of compounds **A**, **B**, and **C**, and explain carefully what each piece of evidence suggests about their identities.

Question 25:

A liquid compound **A** is oily and insoluble in water. However, when heated with aqueous sodium hydroxide, it dissolves. From the reaction mixture, a new liquid **B** can be obtained by distillation.

Liquid **B** reacts with iodine in sodium hydroxide solution, forming a yellow precipitate. When **B** undergoes controlled oxidation, it is converted into compound **C**, which also gives a yellow precipitate with iodine in sodium hydroxide. Additionally, **C** reacts with 2,4-dinitrophenylhydrazine, forming an orange precipitate.

When sulfuric acid is added to the reaction mixture obtained from heating **A** with sodium hydroxide, a white precipitate **D** is formed. **D** reacts with sodium hydrogen carbonate, releasing one mole of carbon dioxide per mole of **D**, indicating a specific functional group.

When **D** is strongly heated with soda-lime, a new compound is formed, which is identified as benzene. Soda-lime is a mixture of sodium hydroxide and calcium oxide that facilitates the removal of a certain functional group through decarboxylation.

Deduce the structures of **A**, **B**, **C**, and **D**, justifying each step logically based on the given experimental observations. Explain how each piece of evidence supports your reasoning.

Question 26:

A water-insoluble oily liquid, **A**, has the molecular formula $C_{10}H_{12}O_2$. When heated with aqueous sodium hydroxide, it dissolves. From the resulting mixture, a volatile liquid **B** can be isolated through distillation.

Liquid **B** reacts with iodine in sodium hydroxide solution, forming a yellow precipitate. When subjected to vigorous oxidation, **B** is transformed into compound **C**, which also produces a yellow precipitate under the same iodine-sodium hydroxide test.

Separately, when sulfuric acid is added to the alkaline reaction mixture of **A** and sodium hydroxide, a white precipitate (**D**) forms. **D** reacts with sodium hydrogen carbonate, releasing carbon dioxide in a 1:1 molar ratio, which provides insight into its molecular structure.

Upon strong heating with soda-lime, **D** is converted into benzene, indicating that a specific functional group has been removed. Soda-lime, a combination of calcium oxide and sodium hydroxide, promotes a process called decarboxylation. This reaction eliminates a carboxyl (-COOH) group and replaces it with a hydrogen atom (-H), reducing the molecular complexity of the compound.

Using the experimental evidence provided, determine the structures of **A**, **B**, **C**, and **D**, and explain in detail how each step supports your deductions.

Question 27:

A water-insoluble, oily liquid **A** has the molecular formula $C_{10}H_{12}O_2$. When heated with aqueous sodium hydroxide, it dissolves, forming a new mixture. From this solution, a volatile liquid **B** can be extracted through distillation.

Unlike some compounds with similar properties, **B** does not produce a yellow precipitate when treated with iodine in sodium hydroxide solution.

Separately, when sulfuric acid is added to the alkaline reaction mixture of **A** and sodium hydroxide, a white precipitate (**C**) forms. This solid reacts with sodium hydrogen carbonate, releasing carbon dioxide in a precise 1:1 molar ratio.

Upon strong heating with soda-lime, **C** undergoes a chemical transformation, yielding benzene. This indicates the removal of a particular functional group through decarboxylation, a process in which a carboxyl (-COOH) group is eliminated and replaced by a hydrogen atom (-H). Soda-lime, a mixture of calcium oxide and sodium hydroxide, facilitates this reaction by creating the necessary conditions for the loss of carbon dioxide (CO_2) and the simplification of the molecular structure.

Analyze the given experimental evidence and deduce the structures of **A**, **B**, and **C**. Provide a detailed explanation of how each observation leads to the correct structural identification.

Question 28:

Compound **A** has a molecular formula of C_4H_8O.

- When sodium metal is added to **A**, effervescence is observed, and the gas produced extinguishes a lighted splint with a 'pop' sound.
- **A** decolorizes aqueous bromine.
- Under reflux, when compound **A** is treated with potassium dichromate(VI), it reacts to produce compound **B** (C_4H_6O), which then gives a positive result when tested with 2,4-dinitrophenylhydrazine.
- When potassium permanganate(VII) is added to **A**, it produces compound **C** ($C_3H_4O_3$) with effervescence.
- **C** forms an orange precipitate with 2,4-dinitrophenylhydrazine.

Deduce the structures of compounds **A**, **B**, and **C**, and explain carefully what each piece of evidence suggests about their identities.

Question 29:

Compound **A** has a molecular formula of C_4H_8O.

- When sodium metal is added to **A**, effervescence is observed, and the gas produced extinguishes a lighted splint with a 'pop' sound.
- **A** decolorizes aqueous bromine.
- Under reflux, when compound **A** is treated with potassium dichromate(VI), it reacts to produce compound **B** ($C_4H_6O_2$).
- When potassium permanganate(VII) is added to **A**, it produces compound **C** ($C_3H_4O_3$) with effervescence.
- **C** forms an orange precipitate with 2,4-dinitrophenylhydrazine.

Deduce the structures of compounds **A**, **B**, and **C**, and explain carefully what each piece of evidence suggests about their identities.

Question 30:

Compound **A** has a molecular formula of C_4H_8O.

- When sodium metal is added to **A**, effervescence is observed, and the gas produced extinguishes a lighted splint with a 'pop' sound.
- **A** decolorizes aqueous bromine.
- Under reflux, when compound **A** is treated with potassium dichromate(VI), it reacts to produce compound **B** ($C_4H_6O_2$).
- When potassium permanganate(VII) is added to **A**, it produces compound **C** ($C_2H_4O_2$) with effervescence.

Deduce the structures of compounds **A**, **B**, and **C**, and explain carefully what each piece of evidence suggests about their identities.

---

11 C, H, O, and X — Clues with Halogens

"In the game of organic chemistry, every bond tells a story; the interplay of carbon, hydrogen, oxygen, and halogens unveils a narrative only the astute can decipher."

— **DT**

In **Chapter 10**, we explored structure elucidation questions involving **carbon, hydrogen, and oxygen**, focusing on interpreting key functional groups like alcohols, carbonyls, and acids. In this chapter, we **add halogens (X)** — and with them, a new set of clues to interpret.

Though halogenoalkanes and halogenoarenes are mostly synthetic, they play a vital role in organic chemistry as intermediates in the production of pharmaceuticals, agrochemicals, and polymers. At the same time, some halogen compounds like **CFCs** serve as a reminder that chemistry has both constructive and destructive power — depending on how the reactivity of halogens is managed.

Why Halogens Matter in Deductive Questions

Halogens introduce new patterns of reactivity, especially in **nucleophilic substitution**. For example, a compound that **"reacts with silver nitrate in ethanol"** likely contains a **halogen ready to leave** — a key clue in many deduction questions.

This chapter also places special emphasis on the **Grignard reaction**, which frequently appears in structure elucidation problems. You'll learn not only what it does, but how to **spot it when it's disguised** as part of a multi-step transformation.

As before, I'll guide you through the thought process step by step — in prose, not tables — showing how to interpret clues, explore pathways, and justify each proposed structure with logic and evidence.

Step-by-Step Complexity: One Element at a Time

We continue our structured progression:

- **Chapter 10**: C, H, and O
- **Chapter 11**: Add halogens (X) → C, H, O, X
- **Chapter 12**: Add nitrogen (N) → C, H, O, X, N

Each added element introduces new functional groups and reactivity. Sometimes, **a single phrase** — like *"gives a white precipitate with silver nitrate in ethanol"* — can entirely shift the correct structure. Recognising and interpreting such cues is at the heart of deduction.

Let's Begin

Halogens bring new twists to familiar logic. This chapter will train you to integrate **halogen reactivity**, substitution pathways, and Grignard chemistry into your deductive toolkit — with confidence and precision.

Let the clues lead the way.

Question 31:

Compound **A** has the chemical formula C_8H_{10}. When exposed to chlorine under ultraviolet light, it reacts to produce compound **B** (C_8H_9Cl). Additionally, **A** undergoes a separate reaction with chlorine under alternative conditions to yield compound **E**, an isomer of **B**.

Compound **B** reacts with aqueous sodium hydroxide to form compound **C** ($C_8H_{10}O$). On the other hand, when **E** is heated with aqueous sodium hydroxide, no reaction takes place. Heating compound **C** with acidified potassium permanganate (VII) results in the formation of compound **D** ($C_7H_6O_2$). Furthermore, reflux of **C** with acidified potassium dichromate (VI) leads to the formation of compound **G**, which reacts with 2,4-dinitrophenylhydrazine to produce an orange precipitate.

When compound **E** is heated with acidified potassium permanganate (VII), effervescence occurs, and compound **F** ($C_7H_5O_2Cl$) is produced. By piecing together this information, deduce the structures of compounds **A** through **G** and explain the logic behind each determination.

Question 32:

Compound **A** has the molecular formula C_8H_{10}. Upon treatment with chlorine under ultraviolet light, it produces compound **B**, which has the formula C_8H_9Cl. Under different reaction conditions, **A** reacts with chlorine again to form compound **E**, an isomer of **B**.

When compound **B** is reacted with aqueous sodium hydroxide, it yields compound **C**, with the formula $C_8H_{10}O$. However, compound **E** does not undergo any reaction when heated with aqueous sodium hydroxide. Heating compound **C** with acidified potassium permanganate (VII) results in the formation of compound **D** ($C_7H_6O_2$). Additionally, reflux of **C** with acidified potassium dichromate (VI) leads to the production of compound **G** ($C_8H_8O_2$).

When compound **E** is heated with acidified potassium permanganate (VII), effervescence is observed, and compound **F** ($C_7H_5O_2Cl$) is formed. Based on this information, deduce the structural formulas of compounds **A** through **G** while providing an explanation for each deduction.

Question 33:

Compound **A** possesses the molecular formula C_8H_{10}. When exposed to chlorine under ultraviolet light, it forms compound **B** with the formula C_8H_9Cl. Additionally, **A** reacts with chlorine under alternate conditions to produce compound **E**, which is isomeric with compound **B**.

Upon treating compound **B** with aqueous sodium hydroxide, compound **C** ($C_8H_{10}O$) is obtained. In contrast, heating compound **E** with aqueous sodium hydroxide results in no reaction. When compound **C** undergoes heating with acidified potassium permanganate (VII), it yields compound **D** ($C_8H_6O_4$). Furthermore, heating compound **C** with acidified potassium dichromate (VI) leads to the formation of compound **G** ($C_8H_8O_2$), and upon strong oxidation, compound **G** is converted back into compound **D**.

When compound **E** is treated with acidified potassium permanganate (VII), decolorization is observed, leading to the formation of compound **F** ($C_8H_5O_4Cl$). Using this information, deduce the structural formulas of compounds **A** through **G** and provide a detailed explanation for each step.

Question 34:

Compound **A**, with the formula $C_4H_8O_2$, reacts with 2,4-dinitrophenylhydrazine to yield an orange precipitate and forms a silver mirror when treated with Tollen's reagent. When treated with acid, compound **A** converts into compound **B** (C_4H_6O), which can exist as either cis or trans isomers. Compound **B** reacts with hydrogen chloride to produce compound **C** (C_4H_7OCl) and undergoes reaction with chlorine to form compound **D** ($C_4H_6OCl_2$).

Upon reduction, compound **C** is transformed into compound **E** (C_4H_9OCl), which, when hydrolyzed using dilute alkali, results in compound **F** ($C_4H_{10}O_2$). Both compounds **A** and **F** give a positive result in the iodoform test (CHI_3), while compound **E** does not react positively. Determine the identities and structures of compounds **A** through **F**, and justify your reasoning.

Question 35:

Compound **A**, with the molecular formula $C_4H_8O_2$, reacts with 2,4-dinitrophenylhydrazine to produce an orange precipitate and forms a silver mirror when treated with Tollen's reagent. On treatment with acid, compound **A** is converted to compound **B** (C_4H_6O), which exists as cis and trans isomers. Compound **B** reacts with hydrogen chloride to form compound **C** (C_4H_7OCl) and reacts with chlorine to give compound **D** ($C_4H_6OCl_2$).

Upon reduction, compound **C** is transformed into compound **E** (C_4H_9OCl), which is hydrolyzed using dilute alkali to yield compound **F** ($C_4H_{10}O_2$). While compound **A** tests positive in the iodoform reaction (CHI_3), compounds **E** and **F** do not. Determine the structural identities of compounds **A** through **F**, providing a detailed explanation for your reasoning.

Question 36:

Compound **A** ($C_4H_8O_2$) produces an orange precipitate when treated with 2,4-dinitrophenylhydrazine and forms a silver mirror upon reaction with Tollen's reagent. On exposure to acid, **A** is converted into compound **B** (C_4H_6O), which does not exhibit stereoisomerism. Compound **B** reacts with hydrogen chloride to produce compound **C** (C_4H_7OCl) and undergoes chlorination to yield compound **D** ($C_4H_6OCl_2$).

Reduction of compound **C** results in the formation of compound **E** (C_4H_9OCl), which is hydrolyzed using dilute alkali to give compound **F** ($C_4H_{10}O_2$). Compound **F** gives a positive result in the iodoform reaction (CHI_3), while compounds **A** and **E** do not. Using this information, deduce the structures of compounds **A** through **F** and provide a detailed explanation of the reasoning behind each step.

Question 37:

Compound **A** ($C_4H_8O_2$) reacts with 2,4-dinitrophenylhydrazine to form an orange precipitate and produces a silver mirror with Tollen's reagent. Upon treatment with acid, **A** transforms into compound **B** (C_4H_6O), which does not exhibit stereoisomerism. Compound **B** reacts with hydrogen chloride to produce compound **C** (C_4H_7OCl) and undergoes chlorination to yield compound **D** ($C_4H_6OCl_2$).

Reduction of compound **C** produces compound **E**, identified as 4-chloro-butan-1-ol, which is then hydrolyzed with dilute alkali to form compound **F** ($C_4H_{10}O_2$). Compounds **A**, **E**, and **F** do not give a positive result in the iodoform test (CHI_3). Using this information, deduce the structures of compounds **A** through **F** and provide a detailed explanation for each step of your reasoning.

Question 38:

Compound **A** ($C_4H_8O_2$) reacts with 2,4-dinitrophenylhydrazine to form an orange precipitate and produces a silver mirror when treated with Tollen's reagent, indicating the presence of an aldehyde functional group. When subjected to acidic conditions, **A** is converted into compound **B** (C_4H_6O), which may exist as cis- and trans-isomers. Compound **B** reacts with hydrogen chloride to yield compound **C** (C_4H_7OCl) and undergoes chlorination to produce compound **D** ($C_4H_6OCl_2$).

Reduction of compound **C** results in the formation of compound **E** (C_4H_9OCl), which is hydrolyzed in the presence of dilute alkali to give compound **F** ($C_4H_{10}O_2$). Compounds **A**, **E**, and **F** do not give a positive result in the iodoform test (CHI_3). Strong oxidation of compound **A** produces compound **G**, with the molecular formula $C_4H_6O_3$. Use this information to deduce the structural identities of compounds **A** through **G**, providing clear reasoning for each step in the process.

Question 39:

Compound **A** ($C_4H_8O_2$) produces an orange precipitate when treated with 2,4-dinitrophenylhydrazine and forms a silver mirror upon reaction with Tollen's reagent. When subjected to acidic conditions, **A** is transformed into compound **B** (C_4H_6O), which can exist as either a cis- or trans-isomer. Compound **B** reacts with hydrogen chloride to give compound **C** (C_4H_7OCl) and undergoes chlorination to produce compound **D** ($C_4H_6OCl_2$).

Reduction of compound **C** results in compound **E** (C_4H_9OCl), which is hydrolyzed by dilute alkali to form compound **F** ($C_4H_{10}O_2$). Compound **F** tests positive in the iodoform reaction (CHI_3), while compounds **A** and **E** do not. Strong oxidation of compound **A** gives compound **G** with the molecular formula $C_4H_6O_3$. Use this information to deduce the identities and structural formulas of compounds **A** through **G**, and explain your reasoning in detail.

Question 40:

Compound **A** (C_3H_7Cl) reacts with aqueous sodium hydroxide, forming compound **C** (C_3H_8O). Treating **C** with phosphorus tribromide results in the release of steamy vapors. Oxidation of **C** using acidified potassium dichromate (VI) produces compound **D** (C_3H_6O), which is further oxidized to compound **E** ($C_3H_6O_2$). Compound **D** reacts positively with 2,4-dinitrophenylhydrazine and Tollen's reagent. Compound **E** reacts with sodium hydrogen carbonate, releasing carbon dioxide, and with phosphorus tribromide.

In the presence of benzene and anhydrous iron(III) chloride, **A** forms compound **B** (C_9H_{12}), whose structure is unexpected based on **A**. Interestingly, compound **B** can also be synthesized from benzene and **F**, an isomer of **A**, to produce the expected product. When treated with sodium hydroxide, **F** is converted into compound **G**, an isomer of **C**. Oxidizing **G** with acidified potassium dichromate (VI) results in compound **H**, an isomer of **D**, with no further oxidation possible. Compound **H** reacts positively with 2,4-dinitrophenylhydrazine, does not react with Tollen's reagent, and gives a positive result in the iodoform test.

Heating **B** with alkaline potassium permanganate (VII) yields an aqueous solution of **I**. Upon treatment with dilute sulfuric acid, **I** produces a white precipitate, identified as **J**.

(a) Identify compounds **A** through **J**, providing detailed explanations of the observations and supporting your reasoning with balanced chemical equations.

(b) Discuss why both **A** and **F** produce the same product when reacted with benzene.

Question 41:

Compound **A** (C_4H_9Cl) undergoes a reaction with aqueous sodium hydroxide to form compound **C** ($C_4H_{10}O$). When treated with phosphorus pentachloride, **C** produces steamy fumes. Oxidation of **C** using acidified potassium dichromate (VI) results in the formation of compound **D** (C_4H_8O), which can be further oxidized to produce compound **E** ($C_4H_8O_2$). Compound **D** reacts positively with both 2,4-dinitrophenylhydrazine and Tollen's reagent. Compound **E**, when treated with sodium hydrogen carbonate solution, releases carbon dioxide and also reacts with phosphorus pentachloride.

In the presence of benzene and anhydrous aluminium chloride, compound **A** reacts to yield compound **B** ($C_{10}H_{14}$), which exhibits a structure differing from what would initially be anticipated based on **A**. Alternatively, compound **B** can be synthesized from benzene and **F**, an isomer of **A**, forming the expected product. Treating **F** with sodium hydroxide produces compound **G**, an isomer of **C**. Oxidizing **G** with acidified potassium dichromate (VI) results in compound **H**, an isomer of **D**, without further oxidation occurring. Compound **H** reacts positively with 2,4-dinitrophenylhydrazine, shows no reaction with Tollen's reagent, and produces a positive result in the iodoform test.

When compound **B** is heated with alkaline potassium permanganate (VII), it generates an aqueous solution of **I**. Upon treatment with dilute sulfuric acid, **I** forms a white precipitate identified as **J**.

(a) Using the provided information, determine the identities and structures of compounds **A** through **J**. Explain each observation comprehensively, supporting your reasoning with balanced chemical equations.

(b) Discuss the reason why both **A** and **F** yield the same product when reacted with benzene.

Question 42:

Compound **A** (C_4H_9Cl) undergoes a reaction with aqueous sodium hydroxide, producing compound **C** ($C_4H_{10}O$). Upon reaction with phosphorus pentachloride, **C** releases steamy fumes. Oxidizing **C** with acidified potassium dichromate (VI) yields compound **D** (C_4H_8O), which can be further oxidized to form compound **E** ($C_4H_8O_2$). Compound **D** exhibits positive responses with both 2,4-dinitrophenylhydrazine and Tollen's reagent, while compound **E** reacts with sodium hydrogen carbonate, releasing carbon dioxide, and also reacts with phosphorus pentachloride.

When reacted with benzene in the presence of anhydrous aluminium chloride, compound **A** forms **B** ($C_{10}H_{14}$), a structure different from what one might predict based on **A**. Compound **B** can alternatively be synthesized from benzene and compound **F**, an isomer of **A**, yielding the anticipated product. Treating **F** with sodium hydroxide produces compound **G**, an isomer of **C**. However, oxidation of **G** with acidified potassium dichromate (VI) is not possible. Additionally, compound **B** is resistant to oxidation by acidified potassium permanganate (VII).

(a) Using the details provided, identify the structures and identities of compounds **A** through **G**, explaining each observation thoroughly and supporting your reasoning with balanced equations.

(b) Offer an explanation as to why both **A** and **G** produce the same product when reacted with benzene.

Question 43:

Compound **A** (C_4H_9Cl) reacts with aqueous sodium hydroxide to produce compound **C** ($C_4H_{10}O$). When treated with phosphorus pentachloride, **C** releases steamy fumes. However, compound **C** does not undergo oxidation when exposed to acidified potassium dichromate (VI).

Compound **A** reacts with benzene in the presence of anhydrous aluminium chloride to yield compound **B** ($C_{10}H_{14}$), which cannot be oxidized by acidified potassium permanganate (VII).

Identify the structures and compositions of compounds **A**, **B**, and **C**, and provide detailed explanations for the observations, including balanced equations to support your reasoning.

Question 44:

Compound **A** ($C_5H_{11}Cl$) reacts with aqueous sodium hydroxide to form compound **B** ($C_5H_{12}O$). When **B** is treated with acidified potassium permanganate (VII), it is converted to compound **C**, which does not produce a yellow precipitate when warmed with alkaline aqueous iodine. Heating **B** with concentrated sulfuric acid at 170°C leads to the formation of compound **D** (C_5H_{10}). Vigorous oxidation of **D** using hot concentrated acidified potassium permanganate (VII) results in the formation of butanoic acid as the sole organic product.

Deduce the structural formulas of compounds **A**, **B**, **C**, and **D**, providing detailed reasoning for each step. Ensure all observations are explained thoroughly, supported by balanced equations.

Question 45:

Compound **A** ($C_5H_{11}Cl$) reacts with aqueous sodium hydroxide to produce compound **B** ($C_5H_{12}O$). Oxidation of **B** using acidified potassium permanganate (VII) forms compound **C**, which yields a yellow precipitate when warmed with alkaline aqueous iodine. Heating **B** with concentrated sulfuric acid at 170°C leads to the formation of two isomers, **D** and **E** (C_5H_{10}). Upon vigorous oxidation using hot, concentrated acidified potassium permanganate (VII), **D** produces both propanoic acid and ethanoic acid as organic products, while **E** exclusively forms butanoic acid.

Deduce the structural formulas of compounds **A**, **B**, **C**, **D**, and **E** based on the observations provided, and explain your reasoning comprehensively, including balanced chemical equations to support your answers.

Question 46:

Compound **A** ($C_5H_{11}Cl$) reacts with aqueous sodium hydroxide, forming compound **B** ($C_5H_{12}O$). Oxidation of **B** using acidified potassium permanganate (VII) results in the formation of compound **C**, which does not produce a yellow precipitate when warmed with alkaline aqueous iodine. When **B** is heated with concentrated sulfuric acid at 170°C, compound **D** (C_5H_{10}) is formed, which may exist as cis- and trans-isomers. Vigorous oxidation of **D** with hot concentrated acidified potassium permanganate (VII) produces propanoic acid and ethanoic acid as the only organic products.

Deduce the structural formulas for compounds **A**, **B**, **C**, and **D**, providing detailed explanations for your conclusions and supporting your reasoning with balanced chemical equations.

Question 47:

Compound **A** ($C_5H_{11}Cl$) reacts with aqueous sodium hydroxide to produce compound **B** ($C_5H_{12}O$). When subjected to oxidation with acidified potassium permanganate (VII), **B** forms compound **C**, which does not produce a yellow precipitate upon warming with alkaline aqueous iodine. Heating **B** with concentrated sulfuric acid at 170°C results in the formation of compound **D** (C_5H_{10}). Vigorous oxidation of **D** with hot concentrated acidified potassium permanganate (VII) exclusively yields butanone as the organic product.

Deduce the structural formulas for compounds **A**, **B**, **C**, and **D** based on the provided information. Provide detailed reasoning for each step and ensure observations are explained thoroughly, supported by balanced chemical equations.

Question 48:

Compound **A** ($C_5H_{11}Cl$) reacts with aqueous sodium hydroxide to yield compound **B** ($C_5H_{12}O$). No observable reaction occurs when compound **B** is treated with acidified potassium permanganate (VII). Heating **B** with concentrated sulfuric acid at 170°C results in the formation of two isomers, **C** and **D** (C_5H_{10}). When subjected to vigorous oxidation using hot concentrated acidified potassium permanganate (VII), **C** produces butanone as the sole organic product, whereas **D** yields ethanoic acid and propanone.

Determine the structural formulas for compounds **A**, **B**, **C**, and **D**, providing a detailed explanation of your reasoning. Ensure observations are clearly justified and supported by balanced chemical equations.

Question 49:

Compound **A** ($C_5H_{11}Cl$) reacts with aqueous sodium hydroxide to form compound **B** ($C_5H_{12}O$). Oxidation of **B** with acidified potassium permanganate (VII) yields compound **C**, which forms a yellow precipitate when warmed with alkaline aqueous iodine. Heating **B** at 170°C with concentrated sulfuric acid produces two isomers, **D** and **E** (C_5H_{10}). When subjected to vigorous oxidation with hot concentrated acidified potassium permanganate (VII), **D** produces 2-methylpropanoic acid as the sole organic product, while **E** results in the formation of ethanoic acid and propanone.

Determine the structural formulas for compounds **A**, **B**, **C**, **D**, and **E** using the given observations. Provide detailed explanations for each step and back up your reasoning with balanced chemical equations.

Question 50:

Compound **A** ($C_5H_{11}Cl$) reacts with aqueous sodium hydroxide, resulting in the formation of compound **B** ($C_5H_{12}O$). Oxidation of **B** using acidified potassium permanganate (VII) produces compound **C**, which does not generate a yellow precipitate when warmed with alkaline aqueous iodine. When **B** is heated with concentrated sulfuric acid at 170°C, it gives rise to compound **D** (C_5H_{10}). Vigorous oxidation of **D** with hot concentrated acidified potassium permanganate (VII) leads to the exclusive formation of 2-methylpropanoic acid as the organic product.

Deduce the structural formulas of compounds **A**, **B**, **C**, and **D** based on the information provided. Ensure that all reasoning is detailed and supported by balanced chemical equations.

Question 51:

Compound **A** ($C_5H_{11}Cl$) undergoes a reaction with aqueous sodium hydroxide, resulting in the formation of compound **B** ($C_5H_{12}O$). Upon oxidation with acidified potassium permanganate (VII), compound **B** is converted to compound **C**. Compound **C** does not yield a yellow precipitate when warmed with alkaline aqueous iodine. When **B** is heated at 170°C in the presence of concentrated sulfuric acid, no compound is produced.

Deduce the structural formulas of compounds **A**, **B**, and **C**, providing a detailed explanation of your reasoning and supporting your conclusions with relevant balanced chemical equations.

Question 52:

Compound **A** (C_2H_4) reacts with hydrogen bromide in the gaseous phase to produce compound **B** (C_2H_5Br). When **B** reacts with aqueous sodium hydroxide, it forms compound **C** (C_2H_6O). Oxidation of **C** with acidified potassium dichromate successively yields compound **D** (C_2H_4O) and then **E** ($C_2H_4O_2$). Compound **D** reacts positively with both 2,4-dinitrophenylhydrazine and Tollen's reagent.

By reacting **B** with magnesium in dry ether, compound **F** (C_2H_5MgBr) is formed. Compound **F** reacts with **D**, yielding compound **G** ($C_4H_{10}O$). Oxidation of **G** with acidified potassium dichromate produces compound **H** (C_4H_8O), which can further react with **F** to form compound **I** ($C_6H_{14}O$). Compound **I** does not undergo oxidation when treated with acidified potassium dichromate.

When compound **A** reacts with bromine in the presence of aqueous sodium chloride, two products are obtained: compound **J** ($C_2H_4Br_2$) and compound **K** (C_2H_4BrCl). Notably, no $C_2H_4Cl_2$ is formed during this reaction.

When **K** is treated with a limited quantity of aqueous sodium hydroxide, compound **L** (C_2H_5OCl) is produced as the major product.

(a) Identify the structural formulas and compositions of compounds **A** through **L** based on the information provided, and explain the observations comprehensively with supporting balanced equations.

(b) Explain why no $C_2H_4Cl_2$ is formed in the reaction described in the second-to-last paragraph.

(c) Justify why C_2H_5OCl is favored over C_2H_5OBr as the major product in the earlier mentioned reaction.

Question 53:

Compound **A** (C_2H_4) undergoes a reaction with hydrogen bromide in the gaseous phase, forming compound **B** (C_2H_5Br). Compound **B** reacts with aqueous sodium hydroxide to yield compound **C** (C_2H_6O), which can then undergo oxidation with acidified potassium dichromate to successively form compound **D** (C_2H_4O) and compound **E** ($C_2H_4O_2$). Compound **D** shows positive reactions with both 2,4-dinitrophenylhydrazine and Tollen's reagent.

Reaction of compound **B** with magnesium in dry ether produces compound **F** (C_2H_5MgBr), which reacts with **D** to give compound **G** ($C_4H_{10}O$). Oxidizing **G** with acidified potassium dichromate produces compound **H** (C_4H_8O), which can further react with **F**, yielding compound **I** ($C_6H_{14}O$). Compound **I** cannot be oxidized when treated with acidified potassium dichromate.

When compound **A** is treated with aqueous bromine, two products are formed: compound **J** (C_2H_4OBr) and compound **K** ($C_2H_4Br_2$). It is notable that compound $C_2H_6O_2$ is not produced in this reaction.

(a) Deduce the identities and structures of compounds **A** through **K**, and provide detailed explanations for each observation, supported by balanced equations where applicable.

(b) Explain the absence of $C_2H_6O_2$ in the reaction involving compound **A** and aqueous bromine.

Question 54:

Compound **A** (C_3H_6) reacts with hydrogen bromide in the gaseous phase, forming compound **B** (C_3H_7Br). Upon reaction with aqueous sodium hydroxide, **B** is converted to compound **C** (C_3H_8O), which can undergo oxidation with acidified potassium dichromate, resulting in compound **D** (C_3H_6O). Compound **D** gives a positive result when tested with 2,4-dinitrophenylhydrazine but does not react with Tollen's reagent.

When treated with magnesium in dry ether, compound **B** produces compound **E** (C_3H_7MgBr). This reagent reacts with **D** to yield compound **F** ($C_6H_{14}O$), which remains unreactive when subjected to oxidation with acidified potassium dichromate.

If compound **A** is treated with aqueous bromine, it produces two compounds: **G** (C_3H_7OBr) and **H** ($C_3H_6Br_2$). Notably, $C_3H_8O_2$ is not formed during this reaction. Compound **G** gives a positive result when tested with alkaline aqueous iodine.

(a) Determine the identities and structures of compounds **A** through **H**, providing detailed explanations for all observations and supporting each step with balanced chemical equations.

(b) Explain why $C_3H_8O_2$ is not produced during the reaction involving compound **A** and aqueous bromine.

Question 55:

Compound **A** (C_3H_6) reacts with hydrogen bromide in the gaseous phase to produce compound **B** (C_3H_7Br). Upon reaction with aqueous sodium hydroxide, compound **B** is converted into compound **C** (C_3H_8O), which can be oxidized using acidified potassium dichromate to form compound **D** (C_3H_6O). Compound **D** reacts positively with 2,4-dinitrophenylhydrazine but does not produce a reaction with Tollen's reagent.

When compound **B** is treated with magnesium in dry ether, compound **E** (C_3H_7MgBr) is formed. This reagent reacts with compound **D**, producing compound **F** ($C_6H_{14}O$), which remains unreactive towards oxidation with acidified potassium dichromate.

Reacting compound **A** with bromine in the presence of sodium chloride results in the formation of two products: compound **G** (C_3H_6BrCl) and compound **H** ($C_3H_6Br_2$). Subjecting compound **G** to a limited reaction with aqueous sodium hydroxide yields compound **I** as the primary product. Compound **I** does not produce a yellow precipitate when tested with alkaline aqueous iodine solution.

Identify the structural formulas and compositions of compounds **A** through **I**, providing detailed explanations for each observation and supporting conclusions with balanced chemical equations where applicable.

Question 56:

Compound **A** (C_4H_8O) reacts with an alkaline aqueous iodine solution and, upon acidification, forms compound **B** ($C_3H_6O_2$). Compound **A** does not react with Tollen's reagent but produces an orange precipitate with 2,4-dinitrophenylhydrazine.

Reduction of compound **A** using lithium aluminium hydride results in the formation of compound **C** ($C_4H_{10}O$). When treated with phosphorus pentachloride, compound **C** gives compound **D** (C_4H_9Cl). Reaction of **D** with magnesium in dry ether produces **H** (C_4H_9MgCl), which reacts with dry ice and undergoes an aqueous acidic work-up to form compound **I** ($C_5H_{10}O_2$).

Compound **I** reacts with compound **C** in the presence of a catalytic amount of concentrated sulfuric acid, yielding compound **J** ($C_9H_{18}O_2$).

When compound **D** is treated with potassium hydroxide in ethanol, it forms three isomeric products with the molecular formula C_4H_8: a pair of geometric isomers, **E** and **F**, along with compound **G**.

Determine the identities and structural formulas of compounds **A** through **J**, thoroughly explaining all observations. Provide detailed reasoning supported by balanced chemical equations.

Question 57:

Compound **A** (C_4H_8O) does not undergo a reaction with alkaline aqueous iodine solution. However, it produces an orange precipitate with 2,4-dinitrophenylhydrazine and reacts positively with Tollen's reagent. Reduction of **A** using lithium aluminium hydride yields compound **B** ($C_4H_{10}O$). Treating **B** with phosphorus pentachloride generates compound **C** (C_4H_9Cl).

When compound **C** reacts with magnesium in dry ether, it forms **F** (C_4H_9MgCl). Compound **F** then reacts with dry ice, followed by an acidic work-up, to produce compound **G** ($C_5H_{10}O_2$). Subsequently, **G** reacts with **B** in the presence of a catalytic amount of concentrated sulfuric acid, resulting in the formation of compound **H** ($C_9H_{18}O_2$).

If compound **C** is treated with potassium hydroxide dissolved in ethanol, compound **D** is formed. Oxidizing **D** with potassium permanganate (VII) results in the formation of compound **E**, accompanied by effervescence. Compound **E** also reacts with sodium hydrogen carbonate, producing additional effervescence.

Identify the structural formulas and compositions of compounds **A** through **H**, thoroughly explaining the observations and providing detailed reasoning supported by balanced equations.

Question 58:

Compound **A** (C_4H_8O) does not undergo a reaction with alkaline aqueous iodine solution**.** However, it reacts positively with 2,4-dinitrophenylhydrazine and Tollen's reagent. Reduction of **A** using lithium aluminium hydride yields compound **B** ($C_4H_{10}O$). Treating **B** with phosphorus pentachloride results in the formation of compound **C** (C_4H_9Cl).

When compound **C** is reacted with magnesium in dry ether, it produces **F** (C_4H_9MgCl). Compound **F** then reacts with dry ice, followed by acidic work-up, to yield compound **G** ($C_5H_{10}O_2$). Compound **G** reacts with **B** in the presence of a catalytic amount of concentrated sulfuric acid, forming compound **H** ($C_9H_{18}O_2$).

When compound **C** is treated with potassium hydroxide dissolved in ethanol, it forms compound **D**. Oxidation of **D** with potassium permanganate (VII) leads to the formation of compound **E**, with effervescence observed. However, compound **E** does not react with sodium hydrogen carbonate.

Identify the structural formulas and compositions for compounds **A** through **H**, providing detailed explanations of the observations and supporting each conclusion with balanced equations.

Question 59:

Compound **A** ($C_5H_{10}O$) reacts with alkaline aqueous iodine solution and, upon acidification, forms compound **B** ($C_4H_8O_2$). Compound **A** does not react with Tollen's reagent but produces an orange precipitate with 2,4-dinitrophenylhydrazine.

Reduction of **A** using lithium aluminium hydride gives compound **C** ($C_5H_{12}O$). Treatment of **C** with phosphorus pentachloride results in compound **D** ($C_5H_{11}Cl$). When **D** is reacted with magnesium in dry ether, it yields compound **H** ($C_5H_{11}MgCl$). Compound **H** then reacts with dry ice, followed by aqueous acid work-up, to produce compound **I** ($C_6H_{12}O_2$).

Compound **I** reacts with compound **C** in the presence of a catalytic amount of concentrated sulfuric acid to form compound **J** ($C_{11}H_{22}O_2$).

When compound **D** is treated with potassium hydroxide in ethanol, three isomers of molecular formula C_5H_{10} are formed: a pair of geometric isomers, **E** and **F**, along with compound **G**.

Determine the identities and structural formulas for compounds **A** through **J**, providing a comprehensive explanation for the observations. Include detailed reasoning supported by balanced chemical equations.

Question 60:

Compound **A**, with the molecular formula $C_5H_{10}O$, reacts with aqueous iodine in the presence of sodium hydroxide. After the reaction mixture is acidified, compound **B** ($C_4H_8O_2$) is obtained. Compound **A** does not react with ammoniacal silver nitrate, but it forms an orange precipitate when treated with 2,4-dinitrophenylhydrazine.

When **A** is treated with lithium aluminium hydride, it is converted into compound **C** ($C_5H_{12}O$). Upon reaction with phosphorus pentachloride, compound **C** gives **D** ($C_5H_{11}Cl$). Compound **D** is then reacted with magnesium in dry ether, forming **G** ($C_5H_{11}MgCl$). This compound reacts with solid carbon dioxide, and after acidic work-up, produces **H** ($C_6H_{12}O_2$).

Compound **H** is reacted with compound **C** in the presence of a catalytic amount of concentrated sulfuric acid, giving compound **I** ($C_{11}H_{22}O_2$).

When compound **D** is treated with potassium hydroxide in ethanol, two isomeric products are formed: **E** and **F**, both with the formula C_5H_{10}. Of these, only **F** produces effervescence when treated with acidified potassium manganate(VII) solution.

Using the information provided, identify the structures of compounds **A** through **I** and explain the observations and transformations involved.

12 C, H, O, X, and N — Mastering the Full Spectrum

"The heart of organic chemistry beats in the harmonious interplay of its constituents; therein lies the challenge, for it is in their subtle interactions that one may discern the true nature of the compound." — **DT**

Welcome to the final stage of your structure elucidation journey — where all five key elements come together: **carbon, hydrogen, oxygen, halogens**, and **nitrogen**. If you've made it this far, you're more than ready for the ultimate test of your reasoning and deduction skills.

In this chapter, you'll tackle problems that go beyond the basics. The presence of nitrogen opens up new structural possibilities — **amines, amides, nitriles**, and more — many of which bring unfamiliar reactivity and hidden clues. You'll need to think carefully, integrate prior knowledge, and stay alert to small details that can shift the entire structure.

Spectroscopy, Logic, and Real-World Thinking

Some questions here will incorporate **^{1}H NMR** and **IR spectroscopy**, tools familiar to **H3 Chemistry students**, but new to many others. Don't worry — you won't need any prior H3 background. Every question includes guiding hints to support your reasoning.

- A **single peak in ^{1}H NMR**? That tells you something powerful about symmetry or the number of proton environments.

- Unfamiliar with **IR** data? Just refer to the absorption bands listed in your **data booklet** — that's exactly what real chemists do.

These problems are not about obscure trivia — they're about **using the right clues to build the right structure**. You're now solving chemistry the way professionals do: with a mix of evidence, logic, and experience.

The Final Step in a Logical Progression

Let's review how we got here:

- **Chapter 10**: Focused on compounds with **C, H, and O**
- **Chapter 11**: Added **halogens (X)** to explore **C, H, O, X** systems
- **Chapter 12**: Introduces **nitrogen (N)** — now the full range: **C, H, O, X, and N**

Each chapter added complexity one element at a time. This final level combines everything, challenging you to apply **chemical tests, DOU, functional group logic**, and now **spectroscopic reasoning**.

Final Functional Group Summary: Your Deduction Cheat Sheet

Now that all core elements are in play, use this reference to anticipate which functional groups might appear — and what clues they tend to leave behind.

Elements Present	**Typical Functional Groups**
C, H	Alkanes, Alkenes, Arenes
C, H, O	Alcohols, Phenols, Ethers, Aldehydes, Ketones, Acids
C, H, X	Halogenoalkanes, Halogenoarenes
C, H, N	Amines, Amides, Nitriles
C, H, O, X	Esters, Acyl chlorides
C, H, O, X, N	Amino acids, peptides, pharmaceuticals, polyfunctional systems

Use this chart as your **field guide** for decoding clues, predicting reactivity, and linking transformations — especially when multiple functional groups are involved.

Ready to Crack the Final Puzzle?

These are the most challenging problems in the book — and the most satisfying. They'll test your ability to connect clues, interpret data, and construct complex structures with confidence.

Take your time, trust your instincts, and let the chemistry speak.

Let's see what you're truly capable of.

Question 61:

Compound **A** (C_4H_9Cl) undergoes a reaction with aqueous sodium hydroxide to produce **B**. When reacted with ammonia, **A** yields **D**. Additionally, treating **A** with potassium cyanide dissolved in ethanol forms **E** (C_5H_9N).

Compound **E** reacts with lithium aluminium hydride in dry ether, resulting in the formation of **F**. When **E** undergoes hydrolysis with aqueous sodium hydroxide followed by treatment with dilute hydrochloric acid, it produces **G** ($C_5H_{10}O_2$). Further reaction of **G** with phosphorus(V) chloride leads to the formation of **H**.

Treating **A** with potassium hydroxide in ethanol leads to the formation of **I**. When compounds **B** and **I** are oxidized using potassium manganate(VII), they yield **C** and **J**, respectively. Compounds **C** ($C_4H_8O_2$) and **J** ($C_3H_6O_2$) both react with sodium hydrogen carbonate to produce effervescence.

Your task is to determine the identities of compounds **A** through **J**, providing detailed reasoning for each step and including the relevant chemical equations.

Question 62:

Compound **A** (C_4H_9Cl) reacts with aqueous sodium hydroxide, yielding **B** ($C_4H_{10}O$). When treated with ammonia, **A** forms **D** ($C_4H_{11}N$). Reaction of **A** with potassium cyanide in ethanol produces **E** (C_5H_9N).

Compound **E** reacts with lithium aluminium hydride in dry ether, resulting in **F** ($C_5H_{13}N$). Additionally, hydrolysis of **E** with aqueous sodium hydroxide followed by treatment with dilute hydrochloric acid leads to **G** ($C_5H_{10}O_2$). Reacting **G** with phosphorus(V) chloride produces **H** (C_5H_9OCl).

If **A** undergoes treatment with potassium hydroxide in ethanol, three isomeric compounds—**J**, **K**, and **I** (all with the molecular formula C_4H_8)—are obtained. Oxidation of **J** and **K** with potassium manganate(VII) yields **M** ($C_2H_4O_2$), while similar treatment of **I** results in **L** ($C_3H_6O_2$). When compound **B** is oxidized under the same conditions, it produces **C** (C_4H_8O).

Compound **C** does not react with ammoniacal silver nitrate but forms an orange precipitate, **N**, upon reacting with 2,4-dinitrophenylhydrazine. Reduction of **N** using tin and excess concentrated hydrochloric acid followed by treatment with aqueous sodium hydroxide yields **O** ($C_{10}H_{16}N_4$).

Determine the identities of compounds **A** through **O**, explaining your reasoning and including relevant chemical equations to justify each step.

Question 63:

Compound **A** (C_4H_9Cl) undergoes a reaction with aqueous sodium hydroxide to produce **B** ($C_4H_{10}O$). When **A** reacts with ammonia, it forms **C** ($C_4H_{11}N$). Additionally, treating **A** with potassium cyanide in ethanol results in the formation of **D** (C_5H_9N).

Compound **D** reacts with lithium aluminium hydride in dry ether to yield **E** ($C_5H_{13}N$). Hydrolysis of **D**, first with aqueous sodium hydroxide followed by treatment with dilute hydrochloric acid, leads to **F** ($C_5H_{10}O_2$). Further reaction of **F** with phosphorus(V) chloride generates **G** (C_5H_9OCl).

When **A** is treated with potassium hydroxide in ethanol, it forms **H**, a compound with the molecular formula C_4H_8. Oxidation of **H** using potassium manganate(VII) results in the formation of **I** (C_3H_6O), with effervescence observed during the reaction. On the other hand, if compound **B** is subjected to the same oxidative conditions, no new product is formed.

Your task is to identify compounds **A** through **I**. For each step, provide detailed reasoning and include the relevant chemical equations to justify your conclusions.

Question 64:

Compound **A** (C_4H_9Cl) reacts with aqueous sodium hydroxide, leading to the formation of **B** ($C_4H_{10}O$). Treating **A** with ammonia produces **D** ($C_4H_{11}N$). Additionally, when **A** is exposed to potassium cyanide dissolved in ethanol, it generates **E** (C_5H_9N).

Compound **E** undergoes reduction with lithium aluminium hydride in dry ether, resulting in **F** ($C_5H_{13}N$). Hydrolysis of **E**, first using aqueous sodium hydroxide followed by dilute hydrochloric acid, gives **G** ($C_5H_{10}O_2$). Subsequent reaction of **G** with phosphorus(V) chloride produces **H** (C_5H_9OCl).

If **A** is treated with potassium hydroxide in ethanol, it forms **I**, a compound with the molecular formula C_4H_8. Oxidation of **I** using potassium manganate(VII) results in **J** (C_3H_6O), accompanied by the observation of effervescence. Under similar oxidative conditions, compound **B** produces a new product, **C**, that can react with sodium metal to generate effervescence and a characteristic 'pop' sound, extinguishing a lighted splint.

Your task is to identify compounds **A** through **J**, providing detailed reasoning for your conclusions and supporting them with relevant chemical equations.

Question 65:

Compound **A** ($C_9H_8O_2$), a solid, effervesces upon contact with sodium bicarbonate solution, releasing carbon dioxide gas. When treated with bromine water, **A** reacts immediately, resulting in decolourisation and the formation of **B** ($C_9H_8O_2Br_2$). Reacting **B** with aqueous sodium hydroxide produces the sodium salt of **C** ($C_9H_{10}O_4$).

Through oxidation using acidified potassium dichromate, **C** is transformed into **D** ($C_9H_6O_4$). Compound **D** shows a positive response to 2,4-dinitrophenylhydrazine but remains unreactive with ammoniacal silver nitrate solution. Upon treatment with phosphorus pentachloride, **D** releases steamy fumes, forming **E** ($C_9H_5O_3Cl$).

Further reaction between **E** and ammonia produces **F** ($C_9H_7O_3N$). All compounds from **A** through **F** undergo oxidation with hot acidified aqueous potassium manganate (VII), resulting in **G** ($C_7H_6O_2$).

Your objective is to identify compounds **A** through **G**. Explain the observations for each step and support your answers with relevant chemical equations.

Question 66:

Compound **A** ($C_{10}H_{12}$) reacts with acidified potassium manganate(VII), causing decolourisation of the solution and resulting in the formation of **B** ($C_8H_6O_4$). When **B** is treated with lithium aluminium hydride in dry ether, it produces **C** ($C_8H_{10}O_2$). Upon treatment with sodium metal followed by chloromethane, **C** is converted into **E** ($C_{10}H_{14}O_2$).

When compound **C** is reacted with phosphorus trichloride, it releases steamy fumes and forms **D** ($C_8H_8Cl_2$). Further treatment of **D** with an excess of ammonia leads to the formation of **F** ($C_8H_{12}N_2$). However, exposing **F** to tin and concentrated hydrochloric acid in excess does not result in the formation of any new product.

Your task is to determine the identities of compounds **A** through **F**, providing detailed explanations for all observations and supporting your conclusions with relevant chemical equations.

Question 67:

Compound **A** ($C_{10}H_8$) reacts with acidified potassium manganate(VII), causing decolourisation of the solution and yielding **B** ($C_8H_6O_4$). When **B** is subjected to heating with soda-lime—a mixture of calcium oxide and sodium hydroxide—it undergoes decarboxylation, producing **C** (C_6H_6).

Compound **C** reacts with chloromethane in the presence of anhydrous iron(III) chloride, resulting in the formation of **D** (C_7H_8). Like **A**, **D** decolourises acidified potassium manganate(VII) solution, forming **E** ($C_7H_6O_2$). Upon treatment with hot concentrated sulfuric acid and nitric acid, **E** undergoes nitration, producing **F** ($C_7H_5NO_3$).

When **F** is reacted with tin and concentrated hydrochloric acid, it is converted to a salt of **G** (C_7H_7N). Compound **G** reacts rapidly with aqueous bromine, decolourising the solution and producing **H** ($C_7H_4O_2NBr_3$).

Identify and explain the observations for compounds **A** through **H**, detailing each transformation step and supporting your conclusions with relevant chemical equations.

Question 68:

Part (a) Compound **A** (C_8H_8) reacts with acidified potassium manganate(VII), leading to the decolourisation of the solution and forming **B** ($C_7H_6O_2$). Additionally, **A** reacts swiftly with liquid bromine, causing the solution to lose its color and resulting in the formation of **C** ($C_8H_8Br_2$).

When **C** is treated with aqueous sodium hydroxide, it undergoes a reaction to produce **D** ($C_8H_{10}O_2$). Upon reflux with acidified sodium dichromate(VI), **D** is transformed into **E** ($C_8H_6O_3$). Compound **E**, however, does not react with ammoniacal silver nitrate or Fehling's solution. It does, however, react with 2,4-dinitrophenylhydrazine, forming an orange-colored compound identified as **F** ($C_{14}H_{10}N_4O_6$).

Explain the observations at each step, detailing the transformations and identifying the functional groups or reactions involved in compounds **A** through **F**.

Part (b) Can potassium manganate(VII) solution be used to convert **D** into **E** instead of using acidified sodium dichromate(VI)? Provide a detailed explanation for your reasoning, considering the reactivity and selectivity of potassium manganate(VII) in oxidation reactions.

Question 69:

Part (a) Compound **A** ($C_{13}H_{12}$) reacts with acidified potassium manganate(VII) solution, causing decolourisation and resulting in the formation of **B** ($C_{13}H_{10}O$). Unlike some compounds, **A** does not decolourise bromine on its own but reacts under specific conditions with bromine to produce another product.

When **A** is exposed to chlorine in the presence of ultraviolet light, it forms **C** ($C_{13}H_{11}Cl$). With an excess of chlorine, **C** is further chlorinated to yield **D** ($C_{13}H_{10}Cl_2$). When **D** undergoes treatment with hot aqueous sodium hydroxide, **B** is generated. However, the structure of **B** differs from what might be anticipated based on the structure of **D**.

Compound **B** shows no reaction with ammoniacal silver nitrate or Fehling's solution. However, **B** reacts with 2,4-dinitrophenylhydrazine, producing an orange-colored derivative, **E** ($C_{19}H_{14}N_4O_4$). When **E** is treated with tin and excess concentrated hydrochloric acid, it is reduced to form a salt of **F** ($C_{19}H_{18}N_4$).

Identify compounds **A** through **F**, providing detailed explanations for the observed reactions and properties at each step. Support your answers with the relevant chemical principles.

Part (b) Draw and describe the structure of the product that would have been expected to form when **D** is treated with hot aqueous sodium hydroxide. Provide an explanation as to why this expected product does not form, incorporating the chemical reasoning behind this unexpected observation.

Question 70:

Compound **A** ($C_{19}H_{16}$) reacts with acidified potassium manganate(VII), leading to decolourisation of the solution and forming **B** ($C_{19}H_{16}O$). Despite this reactivity, **A** does not cause decolourisation of bromine. However, when exposed to bromine under specific conditions, **A** undergoes a reaction to produce a distinct product.

When **A** is treated with excess chlorine in the presence of ultraviolet light, it reacts to form **C** ($C_{19}H_{15}Cl$). Further treatment of **C** with hot alcoholic sodium cyanide leads to the formation of **D** ($C_{20}H_{15}N$). Acidification of **D** results in the generation of **E** ($C_{20}H_{16}O_2$).

Your task is to identify compounds **A** through **E**. For each compound, provide a detailed explanation of the observations, including the chemical transformations and principles that apply to these reactions. Ensure to support your identifications with relevant reasoning and chemical equations.

Question 71:

Compound **A** ($C_{16}H_{23}O_2N$) undergoes hydrogenation in the presence of a nickel catalyst, forming **B** ($C_{16}H_{25}O_2N$). Compound **B** reacts rapidly with bromine in tetrachloromethane, causing the solution to lose its color and producing **C** ($C_{16}H_{24}O_2NBr$). Notably, **C** exists as a pair of optical isomers.

When **A** is oxidised using acidified potassium manganate(VII), it produces three distinct organic products: **E** ($C_8H_7O_4N$), **F** ($C_4H_{10}O$), and **G** (C_3H_6O). Upon reaction with ethanoyl chloride, **E** forms **I** ($C_{10}H_9O_5N$). Effervescence is observed when sodium metal is added to **F**, indicating its chemical reactivity. Compound **G**, when reacted with 2,4-dinitrophenylhydrazine, generates an orange-colored precipitate identified as **H**.

Additionally, when 0.01 mol of **E** reacts with an excess of sodium metal, it liberates 0.24 dm^3 of hydrogen gas at room temperature and pressure. Furthermore, treating **A** with two moles of bromine in liquid form results in the formation of **D** ($C_{16}H_{22}O_2NBr_3$).

Your task is to identify compounds **A** through **I**, providing detailed explanations for all observed reactions and properties. Be sure to support your conclusions with the relevant chemical principles and equations.

Question 72:

Compound **A** (C_3H_6ONCl) is sparingly soluble in water but dissolves when treated with hydrochloric acid, forming **B** ($C_3H_5O_2Cl$). Adding sodium hydrogen carbonate to **B** causes effervescence, indicating the release of carbon dioxide. **B** dissolves in ethanolic potassium hydroxide, producing **C** ($C_3H_4O_2$). However, when **C** is treated with acidified potassium manganate(VII), effervescence is observed, but no organic product is formed.

When **B** is reacted with aqueous potassium hydroxide, followed by acidification, it yields **D** ($C_3H_6O_3$). Upon oxidation with acidified potassium manganate(VII), **D** is converted into **E** ($C_3H_4O_3$). Compound **E** reacts with 2,4-dinitrophenylhydrazine, forming an orange-yellow compound identified as **F** ($C_9H_8O_6N_4$). Further reaction of **F** with tin and excess concentrated hydrochloric acid, followed by alkaline hydrolysis, results in the formation of **G** ($C_9H_{12}O_2N_4$).

Your task is to determine the identities of **A** through **G**, providing comprehensive explanations for all observed reactions and transformations. Ensure to support your conclusions with detailed reasoning and appropriate chemical equations.

Question 73:

Compound **A** (C_4H_8ONCl) is only sparingly soluble in water but dissolves readily in hot aqueous sodium hydroxide, to yield two products: **B** ($C_2H_4O_2$) and **C** (C_2H_7NO). When sodium hydrogen carbonate is added to **B**, effervescence occurs, indicating the release of carbon dioxide gas.

Compound **C**, when treated with acidified potassium manganate(VII), undergoes oxidation, resulting in the formation of **D** ($C_2H_5NO_2$). Notably, **D** exists as a dipolar species, reflecting its zwitterionic structure.

Your task is to identify compounds **A** through **D**, providing detailed explanations for all observed chemical reactions, physical properties, and transformations. Be sure to support your identifications with sound chemical reasoning and balanced equations where appropriate.

Question 74:

Compound **A** (C_6H_{12}) shows a single peak in its ^1H NMR spectrum, which indicates that all hydrogens in the molecule are chemically equivalent due to its symmetrical structure. **A** decolorizes bromine water. In a separate procedure, refluxing **A** with acidified potassium permanganate oxidizes it to give compound **B**.

When **B** is treated with 2,4-dinitrophenylhydrazine, it produces an orange-colored derivative identified as **C** ($C_9H_{10}N_4O_4$). Reacting **C** with tin and excess concentrated hydrochloric acid, followed by subsequent alkaline hydrolysis, converts it into **D** ($C_9H_{14}N_4$).

Additionally, when **B** is reacted with bromoethane in the presence of magnesium in dry ether, it results in the formation of **E** ($C_5H_{12}O$). Compound **E** does not react with acidified potassium manganate(VII), as no decolourisation occurs.

Additional Insight:

In ^1H NMR spectroscopy, the presence of a single peak is a unique feature observed in molecules where all hydrogen atoms are equivalent due to molecular symmetry. Examples of such molecules include methane (CH_4) and tetramethylsilane ($Si(CH_3)_4$), where identical environments produce a single resonance signal.

Your task is to identify compounds **A** through **E** and explain all observations related to their reactions and properties, incorporating chemical principles and balanced equations to support your reasoning.

Question 75:

Compound **A** (C_6H_{12}) exhibits a single peak in its ^1H NMR spectrum, which indicates that all hydrogen atoms are in an identical chemical environment. This suggests a high degree of molecular symmetry. **A** does not decolourise bromine water, showing a lack of unsaturation. However, when exposed to chlorine in the presence of ultraviolet light, it forms **B** ($C_6H_{11}Cl$).

Compound **B** reacts with aqueous sodium hydroxide, leading to the formation of **C** ($C_6H_{12}O$). When **C** is heated with acidified potassium manganate(VII), it undergoes oxidation to produce **D**, a compound that gives an orange precipitate with 2,4-dinitrophenylhydrazine.

Reaction of **D** with bromoethane in the presence of magnesium in dry ether leads to the formation of **E** ($C_8H_{16}O$), which does not react with acidified potassium manganate(VII). Passing hydrogen cyanide gas through **D** results in the formation of **F** ($C_7H_{11}ON$), which, when acidified, converts to **G** ($C_7H_{12}O_3$).

When **B** is heated with alcoholic sodium hydroxide, **H** (C_6H_{10}) is generated. Oxidation of **H** with acidified potassium manganate(VII) yields **I** ($C_6H_{10}O_4$). Treating **I** with phosphorus pentachloride results in the release of steamy fumes and the formation of **J** ($C_6H_8Cl_2O_2$). Finally, when **J** is treated with ammonia, it reacts to form **K** ($C_6H_{12}O_2N_2$).

Your task is to identify compounds **A** through **K**, providing detailed explanations for each observed reaction and transformation. Be sure to support your answers with logical reasoning and balanced chemical equations.

Question 76:

Compound **A** ($C_3H_6Cl_2$) exhibits optical activity, implying the presence of a chiral center in its structure. When treated with aqueous sodium hydroxide, **A** undergoes a reaction that results in the formation of **B** ($C_3H_8O_2$). Heating **B** with acidified potassium manganate(VII) causes an oxidation reaction, giving rise to **C** ($C_3H_4O_3$).

Using sodium borohydride in pyridine to reduce **C** produces **D** ($C_3H_6O_3$). When **D** is subjected to concentrated sulfuric acid at 170°C, dehydration occurs, leading to the formation of **E** ($C_3H_4O_2$). The reduction of **E** with lithium aluminium hydride in dry ether transforms it into **F** (C_3H_6O).

Further oxidation of **F** with acidified potassium dichromate(VI) with immediate distillation results in **G** (C_3H_4O). Compound **G**, when exposed to hydrogen cyanide gas in the presence of trace sodium hydroxide, undergoes a reaction to form **H** (C_4H_5ON). Further treatment of **H** with anhydrous lithium aluminium hydride in dry ether yields **I** ($C_4H_{11}ON$).

Your task is to identify compounds **A** through **I**, giving detailed explanations of the processes and transformations involved. Include balanced chemical equations and scientific reasoning to support your answers.

Question 77:

Compound **A** ($C_3H_6Cl_2$) lacks enantiomerism, indicating the molecule is not chiral and does not have non-superimposable mirror images. When **A** reacts with aqueous sodium hydroxide, the expected product **B** ($C_3H_8O_2$) is surprisingly not formed. Instead, the reaction produces **C** (C_3H_6O), which undergoes positive tests with 2,4-dinitrophenylhydrazine, ammoniacal silver nitrate, and Fehling's solution.

In the presence of an acid catalyst, two molecules of **C** combine, releasing one molecule of water, and forming **D** ($C_6H_{10}O$). Oxidation of **D** with acidified sodium dichromate(VI) yields **E** ($C_6H_{10}O_2$). When **D** is instead treated with hot acidified potassium manganate(VII), it is oxidized to form **F** ($C_3H_4O_3$), along with a secondary product, propanoic acid.

Further examination of **F** reveals its reaction with warm alkaline aqueous iodine. Upon acidification, the mixture produces the organic compound **G**, accompanied by the formation of CHI_3 (iodoform). Additionally, **F** reacts with 2,4-dinitrophenylhydrazine, producing an orange precipitate identified as **H** ($C_9H_8O_6N_4$).

Your task is to identify the structures of **A** through **H**. Provide detailed explanations for all reactions and observations, incorporating chemical reasoning and balanced equations to substantiate your conclusions.

Question 78:

Compound **A** ($C_3H_6Cl_2$) is a symmetrical molecule that lacks chirality, meaning it is non-enantiomeric. When reacted with aqueous sodium hydroxide, A undergoes substitution and hydrolysis, forming **B** ($C_3H_8O_2$). Oxidizing **B** with acidified potassium manganate(VII) results in **C**, identified as $C_3H_4O_4$, a highly oxidized derivative.

If **B** is treated with acidified potassium dichromate(VI) and the reaction is distilled promptly, the product **D** ($C_3H_4O_2$) is obtained. Introducing cold hydrogen cyanide gas to **D** in the presence of trace sodium hydroxide induces nucleophilic addition, generating **E** ($C_5H_6N_2O_2$). Dissolving **E** in hot hydrochloric acid produces **F** ($C_5H_8O_6$), which undergoes reduction when treated with lithium aluminium hydride in dry ether, yielding **G** ($C_5H_{12}O_4$).

Remarkably, when sodium metal is added to **G**, the reaction releases twice the amount of hydrogen gas as there are moles of **G**, highlighting the molecular structure's reactive hydrogen atoms. Separately, reacting **A** with ethanolic potassium cyanide results in **H** ($C_5H_6N_2$), which, upon acidification, transforms into **I** ($C_5H_8O_4$).

Your challenge is to uncover the identities of compounds **A** through **I** while detailing the observations and chemical principles involved in each reaction. Ensure your answers are supported by logical deductions and balanced chemical equations.

Question 79:

Compound **A** ($C_3H_6Cl_2$) is non-chiral, confirming it does not exhibit optical isomerism. Treatment of **A** with aqueous sodium hydroxide does not yield the anticipated product **B** ($C_3H_8O_2$). Instead, the reaction produces **C** (C_3H_6O). Testing **C** with 2,4-dinitrophenylhydrazine gives a positive result, but it does not react with ammoniacal silver nitrate or Fehling's solution.

In the presence of an acid catalyst, two molecules of **C** condense with the loss of water to form **D** ($C_6H_{10}O$). When **D** is treated with hot acidified potassium dichromate(VI), there is no new compound produced, suggesting resistance to this oxidizing agent. However, when heated with acidified potassium manganate(VII), **D** undergoes oxidation to produce **E** ($C_3H_4O_3$), along with the regeneration of **C** as a by-product.

When **E** is treated with alkaline aqueous iodine, the reaction leads to the formation of **F**, an organic salt, accompanied by the production of CHI_3 (iodoform). Additionally, **E** reacts with phosphorus tribromide to form **G** ($C_3H_3BrO_2$). Further treatment of **G** with ammonia results in the formation of **H** ($C_3H_5NO_2$).

Your task is to identify the molecular structures of **A** through **H**, providing detailed explanations for each reaction. Ensure your reasoning is supported by clear chemical principles and balanced equations where applicable.

Question 80:

Compound **A** (C_6H_{12}) produces a single peak in its ^1H NMR spectrum, which reflects the uniform chemical environment of all hydrogen atoms due to the molecular symmetry. **A** does not react with bromine water, indicating the absence of unsaturated bonds. Subjecting **A** to chlorine exposure under ultraviolet light produces **B** ($C_6H_{11}Cl$). When **B** reacts with aqueous sodium hydroxide under reflux conditions, it yields **C** ($C_6H_{12}O$), an alcohol derivative.

Heating **C** with acidified potassium manganate(VII) oxidizes it to form **D** ($C_6H_{10}O$). Treating **D** in the presence of sodium ethanoate and then acidifying the reaction mixture causes two molecules of **D** to combine, producing **E** ($C_{12}H_{18}O$), while eliminating one molecule of water. When heated with acidified potassium manganate(VII), **E** yields **F** ($C_6H_8O_2$) and regenerates **D** as a secondary product.

Hydrogen cyanide gas is bubbled through **F** in the presence of trace sodium hydroxide, resulting in the formation of **G** ($C_8H_{10}N_2O_2$). Acidification of **G** transforms it into **H**. Finally, **G** is subjected to hydrogen gas under elevated pressure in the presence of a nickel catalyst, reducing it to **I**.

Your task is to determine the identities of compounds **A** through **I**, providing detailed explanations for the observations recorded at each stage of the process. Include balanced chemical equations and logical reasoning to support your conclusions.

Question 81:

Compound **A** (C_8H_{10}) exhibits two distinct peaks in its ^{1}H NMR spectrum, indicating that its hydrogen atoms are distributed in two chemically different environments. Reaction of **A** with potassium manganate(VII) results in the formation of **B** ($C_8H_6O_4$).

Treatment of **A** with ethanoyl chloride in the presence of anhydrous aluminum chloride produces **C** ($C_{10}H_{12}O$). Analysis of **C** using IR spectroscopy reveals a strong absorption band at 1685 cm^{-1}. Reduction of **C** with sodium borohydride in pyridine yields **D** ($C_{10}H_{14}O$), which exists in stereoisomeric forms. IR spectroscopy of **D** shows a broad absorption band at 3340 cm^{-1}.

Heating **D** with concentrated sulfuric acid results in the elimination of a molecule, producing **E**, which displays an absorption band near 1630 cm^{-1} in its IR spectrum. When **C** is reacted in the presence of sodium ethanoate, followed by acidification, two molecules of **C** combine with the loss of a water molecule to form **F** ($C_{20}H_{22}O$).

Treating **F** with bromoethane and magnesium in dry ether results in the formation of **G** ($C_{22}H_{28}O$). Additionally, bubbling hydrogen cyanide gas in the presence of a catalytic amount of NaOH through **F** leads to the generation of **H** ($C_{21}H_{23}NO$).

Your task is to identify compounds **A** through **H**, providing detailed explanations for the observations recorded during each transformation. Ensure to support your reasoning with balanced chemical equations and logical deductions.

Part 4

Case Files — Solutions & Strategies

Answer key with simplified flowcharts and thought process explanations.

13 Answers to Chapter 10 — C, H, and O

In this chapter, I walk you through my thought process for solving the structure elucidation questions from **Chapter 10: C, H, and O — A Systematic Start**.

Each solution includes a **simplified flowchart** showing the correct structures — your visual **answer key**. These flowcharts are intentionally **incomplete**, omitting some reagents and conditions. This is deliberate: it challenges you to fill in the missing details and strengthens your ability to deduce the steps needed to transform one compound into another.

Use the flowcharts to confirm your structures — and go one step further by reconstructing the full synthetic path with reagents and conditions, just as a top scorer would.

Question 1:

Compound **A** has the molecular formula C_4H_8. When treated with cold concentrated sulfuric acid, it yields compound **B**, which has the formula $C_4H_{10}O$. Compound **B** subsequently undergoes oxidation when reacted with acidified sodium dichromate for prolonged period to produce compound **C**, C_4H_8O. Additionally, compound **A** is capable of decolorizing liquid bromine. Furthermore, upon reaction with hot acidified potassium permanganate(VII), compound **A** generates compound **D**, $C_3H_6O_2$.

Propose structures for compounds **A**, **B**, **C** and **D**, and discuss the observations made during these reactions.

My thought process:

Compound **A** exhibits one degree of unsaturation, and its ability to decolorize bromine confirms that it is an alkene. Since C_4H_8 can exist in two isomeric forms, the key to distinguishing the correct structure lies with the observations concerning compound **D**. The fact that **D**, formed after the oxidative cleavage of **A**, also has one degree of unsaturation and contains one carbon less than **A** suggests that **D** is propanoic acid, which in turn implies that **A** is a terminal alkene.

When **A** undergoes hydration to form alcohol **B**, the reaction does not necessarily follow Markovnikov's rule in the context of structure determination. This means we cannot assume that the hydroxyl group will always attach to the more substituted carbon. Knowing that **B** is an alcohol and can be oxidized by sodium dichromate, it must be either a primary or secondary alcohol. If **B** were primary, its oxidation under prolonged conditions with a mild oxidizing agent would lead to a carboxylic acid rather than an aldehyde. However, because compound **C** is identified as a ketone, this indicates that the alcohol **B** must be secondary.

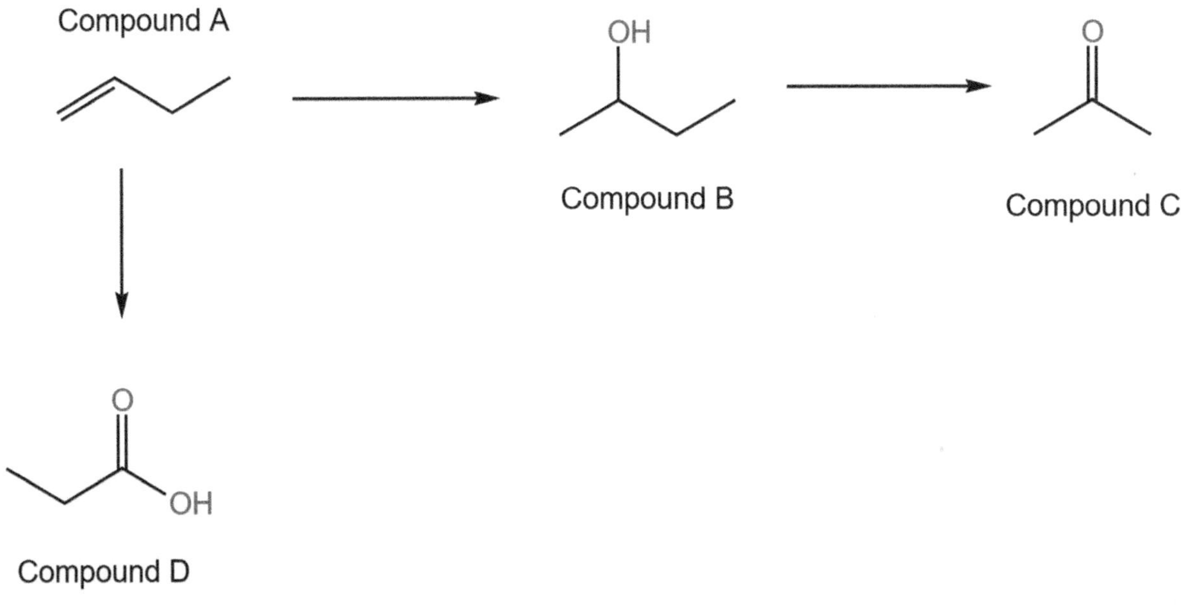

Question 2:

Compound **A** has a molecular formula of C_5H_{10} and is known to decolorize aqueous bromine. When compound **A** reacts with acidified potassium permanganate(VII), it yields compound **B**, which has the formula C_4H_8O. Furthermore, compound **B** reacts with 2,4-dinitrophenylhydrazine, resulting in the formation of an orange precipitate.

Propose structures for compounds **A**, **B**, and discuss the observations made during these reactions.

My thought process:

Compound **A** is an alkene because it decolorizes aqueous bromine. When subjected to oxidative cleavage with $KMnO_4$, it produces compound **B**, which has one carbon fewer than **A**. This observation implies that **A** must be a terminal alkene. Additionally, under these conditions, the oxidizing agent $KMnO_4$ yields a ketone rather than an aldehyde, carboxylic acid, or carbon dioxide. For **B** to be a four-carbon ketone, the only possibility is butanone. Consequently, if **B** is butanone and **A** is a terminal alkene, then **A** must be 2-methylbut-1-ene.

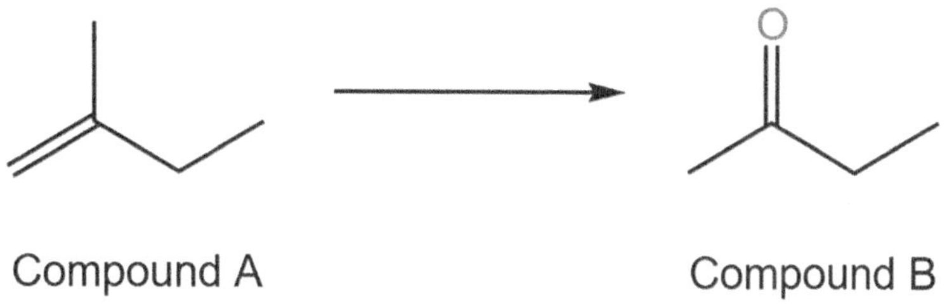

Question 3:

Compound **A** has a molecular formula of C_5H_{10} and is capable of decolorizing liquid bromine. Compound **A** also reacts with potassium permanganate(VII) to produce compound **B**, which has the formula $C_4H_8O_2$, while releasing a gas that forms a white precipitate with calcium hydroxide. It is noteworthy that compound **B** does not yield an orange precipitate when treated with Brady's reagent. Additionally, compound **B** can be reduced by lithium aluminium hydride ($LiAlH_4$) to form compound **C**, which has the formula $C_4H_{10}O$. Upon treatment with hot concentrated sulfuric acid, compound **C** yields compound **D** (C_4H_8). When subjected to oxidation, compound **D** produces compound **E** ($C_3H_6O_2$), which generates effervescence upon exposure to sodium hydrogen carbonate.

Propose structures for compounds **A** through **E** and elucidate the observations described in the reactions.

My thought process:

It's important not to assume that you have to work through every statement sequentially when solving a structure elucidation problem. In this case, it's more straightforward to start with compound **E** because its smaller number of carbon atoms means there are fewer possible isomers. Given that **E** has a degree of unsaturation of 1 and contains two carbon atoms, it is most likely a carboxylic acid—specifically, propanoic acid.

Since compound **D** has one more carbon atom than **E**, it follows that **D** is a terminal alkene, which would be but-1-ene. Moving on to compound **C**, we know it is an alcohol, but its exact classification as primary or secondary isn't immediately clear. This ambiguity arises because the dehydration of alcohol **C** produces alkene **D**, and there are two possible ways to add hydrogen and hydroxyl groups across the double bond.

🔍 **Tip for students:** It is crucial for students to avoid applying Markovnikov's or Zaitsev's rules when predicting structures in these problems. While these rules are useful for predicting major products in organic synthesis, structure elucidation relies on piecing together clues from the question itself.

Additionally, compound **B** is identified as a carboxylic acid. When reduced with $LiAlH_4$, it would yield a primary alcohol. This information suggests that compound **C** must be a primary alcohol with four carbon atoms, which corresponds to butan-1-ol. Consequently, if **C** is butan-1-ol, then **B** must be butanoic acid. Since **B** has one fewer carbon atom than **A**, compound **A** must be a terminal alkene, meaning **A** is pent-1-ene.

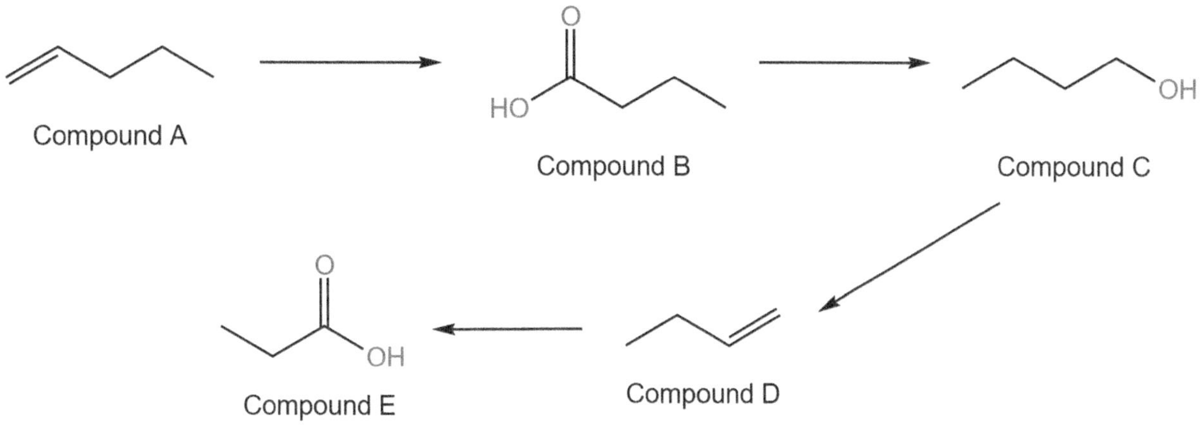
HO
O
OH
OH
O
Compound A
Compound B
Compound C
Compound D
Compound E

Question 4:

Compound **A** has a molecular formula of C_5H_{10} and can decolorize aqueous bromine. Upon treatment with potassium permanganate(VII), it produces compound **B**, which has the formula $C_4H_8O_2$. There is no observable reaction between compound **B** and 2,4-dinitrophenylhydrazine. When lithium aluminium hydride is added to compound **B**, it yields compound **C**, with the formula $C_4H_{10}O$. The addition of excess hot concentrated sulfuric acid to compound **C** results in the formation of compound **D,** C_4H_8.

When subjected to strong oxidation, compound **D** generates compound **E** along with effervescence. Additionally, compound **E** forms an orange precipitate when treated with Brady's reagent.

Suggest structures for compounds **A** through **E** and explain the observations described above.

My thought process:

Every structure elucidation question has a weak point that can be exploited, and in this case, the key lies with compound **E**—even though its molecular formula isn't provided. I began by examining compound **D**, a four-carbon alkene. Considering the possibilities—namely but-1-ene, but-2-ene (both cis and trans), and 2-methylpropene—the production of effervescence during strong oxidation suggests that **D** must be a terminal alkene. This observation rules out but-2-ene. Furthermore, only 2-methylpropene would yield a ketone (compound **E**) upon oxidative cleavage, as evidenced by its reaction with Brady's reagent (2,4-dinitrophenylhydrazine), whereas but-1-ene would produce a carboxylic acid. Consequently, compound **E**, being a three-carbon ketone, is identified as propanone, and compound **D** is deduced to be 2-methylpropene.

Next, I considered compound **B**, which is a carboxylic acid containing four carbons. Upon reduction with $LiAlH_4$, it forms compound **C**, a primary alcohol. Knowing that **D** is 2-methylpropene, the addition of H and OH must occur in such a way that **C** is a primary alcohol; hence, **C** is 2-methylpropan-1-ol. This indicates that compound **B** is 2-methylpropanoic acid. Finally, since **B** has one less carbon atom than compound **A** (an alkene), **A** must be a terminal alkene with an extra carbon, making it 3-methylbut-1-ene.

This approach demonstrates a reverse-engineering method where I used key reaction clues to narrow down the possibilities and systematically determine the structures of compounds **A** through **E**.

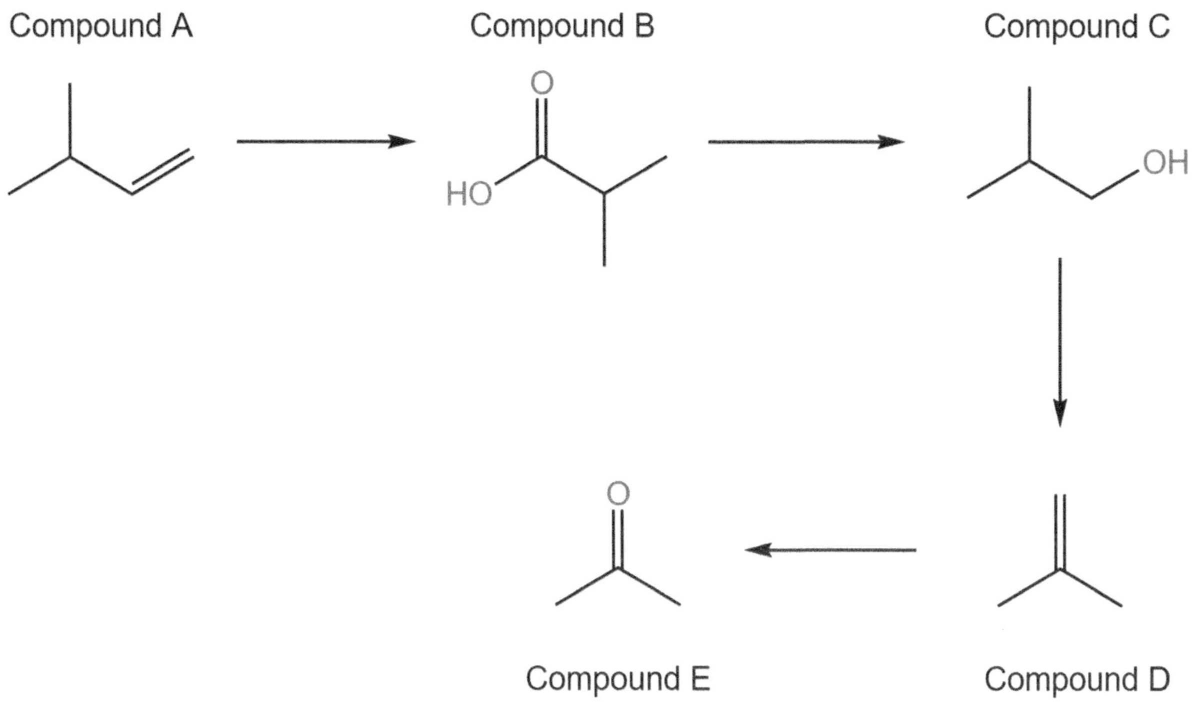

Compound A
Compound B
Compound C
Compound D
Compound E
O
HO
OH
O

Question 5:

Compound **A** has a molecular formula of C_5H_{10}, which is capable of decolorizing bromine water. When subjected to acidified potassium permanganate(VII), it produces two organic compounds: compound **B** with the formula $C_2H_4O_2$ and compound **C** with the formula $C_3H_6O_2$. Notably, both compounds **B** and **C** do not exhibit any observable change when treated with 2,4-dinitrophenylhydrazine.

Suggest structures for compounds **A**, **B**, and **C**, and provide an explanation for the observations described above.

My thought process:

Compound **A** is a five-carbon alkene, as indicated by its ability to decolorize bromine water. When oxidatively cleaved by acidified $KMnO_4$, **A** splits into two fragments—compound **B** ($C_2H_4O_2$) and compound **C** ($C_3H_6O_2$). The total number of carbon atoms in **B** and **C** equals that in **A**, which suggests that **A** is not a terminal alkene. Furthermore, both **B** and **C** are unreactive toward 2,4-dinitrophenylhydrazine, indicating that they are carboxylic acids rather than aldehydes or ketones.

Since **B** contains two carbon atoms and is a carboxylic acid, it must be ethanoic acid. Similarly, **C**, with its three carbon atoms, can only be propanoic acid. By piecing together these fragments, the structure of **A** can be deduced as pent-2-ene.

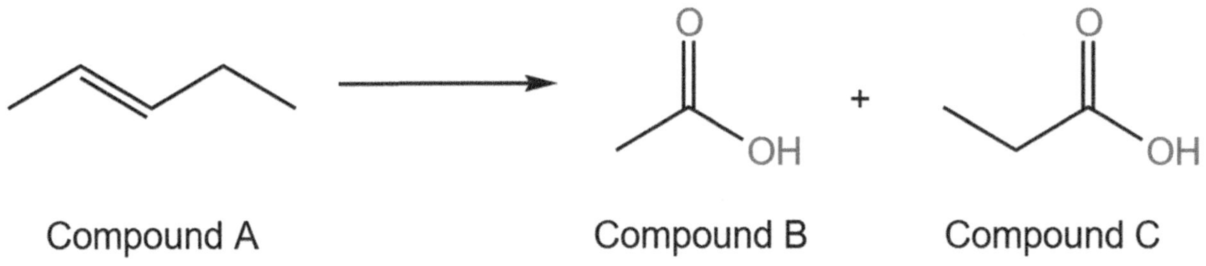

Question 6:

Compound **A** has the molecular formula C_5H_{10}, which decolorizes bromine water upon treatment. When subjected to acidified potassium permanganate(VII), it yields two organic compounds: compound **B** with the formula C_3H_6O and compound **C** with the formula $C_2H_4O_2$. Only compound **B** forms an orange precipitate when treated with Brady's reagent.

Suggest structures for compounds **A**, **B**, and **C**, and provide a detailed explanation for the observations described above.

My thought process:

I began by focusing on the oxidative cleavage products. Compound **A** (C_5H_{10}) reacts with acidified $KMnO_4$ to yield two fragments—compound **B** (C_3H_6O) and compound **C** ($C_2H_4O_2$)—whose combined carbon count equals that of A. This indicates that **A** is an internal (non-terminal) alkene. Under these conditions, $KMnO_4$ converts alkenes into ketones, carboxylic acids, or carbon dioxide; it does not produce aldehydes. The fact that compound **C** does not react with 2,4-dinitrophenylhydrazine (Brady's reagent) tells us that it is a carboxylic acid, and with only two carbons available, it must be ethanoic acid. Meanwhile, compound **B** gives a positive Brady's test, confirming it is a ketone, and since it contains three carbons, the only possibility is propanone. Piecing these fragments together leads to the conclusion that compound **A** is 2-methylbut-2-ene.

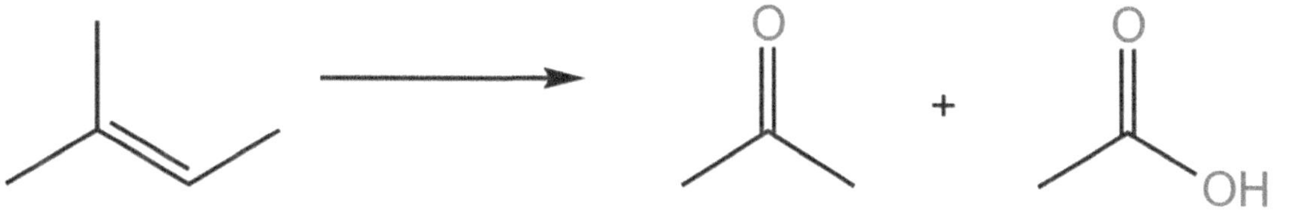

Compound A Compound B Compound C

Question 7:

Compound **A** has the molecular formula C_3H_8O. It reacts with hot concentrated phosphoric acid to yield compound **B**, which has the formula C_3H_6 and rapidly decolorizes a solution of bromine in tetrachloromethane. When treated with ethanoyl chloride, compound **A** produces compound **C**, with the molecular formula $C_5H_{10}O_2$. Additionally, when compound **A** is oxidized using potassium dichromate in dilute sulfuric acid, it forms compound **D**, C_3H_6O, which gives an orange precipitate with 2,4-dinitrophenylhydrazine but does not react with an ammoniacal solution of silver nitrate.

Deduce the structures of compounds **A**, **B**, **C**, and **D**, and explain carefully what each piece of evidence tells you about these compounds.

My thought process:

My reasoning begins by noting that compound **A**, with the formula C_3H_8O, is a three-carbon alcohol. Although many texts use hot concentrated sulfuric acid for dehydration, here hot concentrated phosphoric acid is employed—a reminder that alternative reagents can achieve the same result. At this point, **A** could be either propan-1-ol or propan-2-ol. The dehydration of **A** produces compound **B** (C_3H_6), an alkene, which is consistent with the elimination of water from an alcohol.

Next, when compound **A** reacts with ethanoyl chloride, it forms compound **C** ($C_5H_{10}O_2$). This reaction is a condensation between the alcohol and the acyl chloride, where the -OH group of **A** and the chlorine atom from ethanoyl chloride are eliminated as HCl, leading to the formation of an ester. The degree of unsaturation in compound **C** further confirms its ester structure.

The key evidence comes from the oxidation of compound **A** using potassium dichromate in dilute sulfuric acid, which yields compound **D** (C_3H_6O). Compound **D** gives an orange precipitate with 2,4-dinitrophenylhydrazine, indicating the presence of a carbonyl group, but it does not react with ammoniacal silver nitrate (Tollens' reagent). This behavior confirms that **D** is a ketone rather than an aldehyde, because aldehydes typically react with Tollens' reagent. Given that **D** is a three-carbon ketone, it can only be propanone. This finding rules out the possibility of **A** being a primary alcohol (propan-1-ol), since oxidation of a primary alcohol would yield an aldehyde and eventually a carboxylic acid. Therefore, **A** must be a secondary alcohol—specifically, propan-2-ol.

Consequently, the dehydration of propan-2-ol forms propene (compound **B**), its reaction with ethanoyl chloride produces an ester (compound **C**), and its oxidation yields propanone (compound **D**), thereby conclusively identifying compound **A** as propan-2-ol.

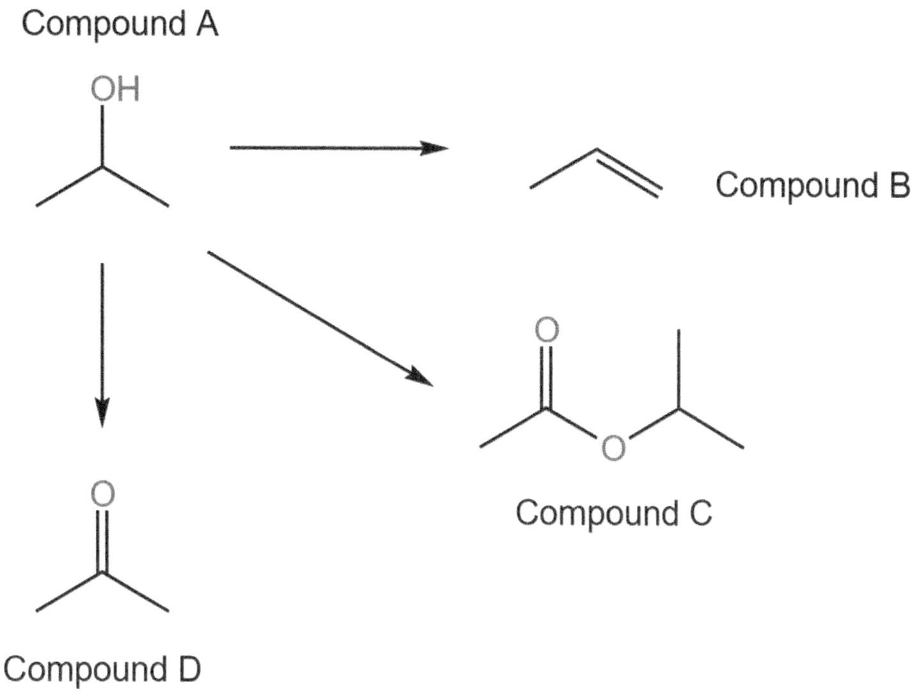
Compound A
OH
Compound B
O
O
Compound C
O
Compound D

Question 8:

Compound **A** has the molecular formula C_3H_8O. It reacts with hot concentrated phosphoric acid to yield compound **B**, which has the formula C_3H_6 and rapidly decolorizes a solution of bromine in tetrachloromethane. When treated with ethanoyl chloride, compound **A** produces compound **C**, with the molecular formula $C_5H_{10}O_2$. Additionally, when compound **A** is oxidized using potassium dichromate in dilute sulfuric acid, it forms compound **D** ($C_3H_6O_2$), which gives effervescence with potassium hydrogen carbonate.

Deduce the structures of compounds **A**, **B**, **C**, and **D**, and explain carefully what each piece of evidence tells you about these compounds.

My thought process:

This problem is very similar to Q7, with the main difference being the formula for compound **D**. Compound **D** ($C_3H_6O_2$) has a degree of unsaturation of one and contains two oxygen atoms. Since it is an oxidized product, these characteristics indicate that **D** is propanoic acid, as it has three carbon atoms and fits the transformation of an alcohol to a carboxylic acid. This outcome means that compound **A** must be a primary alcohol—specifically, propan-1-ol—because only a primary alcohol can oxidize to form a carboxylic acid. When **A** is dehydrated using hot concentrated phosphoric acid, it produces compound **B**, which is propene. Additionally, when **A** reacts with ethanoyl chloride, it yields compound **C** ($C_5H_{10}O_2$), an ester formed by removing a hydrogen atom from the alcohol and a chlorine atom from ethanoyl chloride, thereby linking the two fragments via an ester bond.

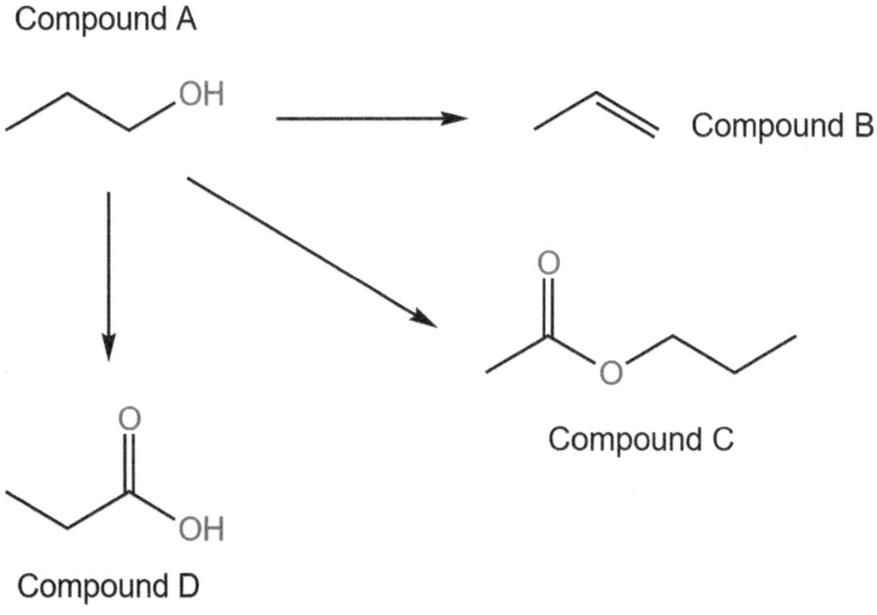

Question 9:

Compound **A** ($C_4H_{10}O$) does not react with potassium dichromate(VI). When heated with concentrated sulfuric acid, it forms compound **B** (C_4H_8), which causes the decolorization of aqueous bromine.

Compound **B** reacts with acidified potassium permanganate(VII) to produce compound **C** (C_3H_6O), accompanied by the evolution of a gas that forms a white precipitate when bubbled through calcium hydroxide solution.

Deduce the structures of compounds **A**, **B**, and **C**, and explain carefully what each piece of evidence suggests about their identities.

My thought process:

My approach begins by focusing on compound **C**, which is the key to unraveling the problem. Since compound **B** is an alkene, compound **C** is produced by its oxidative cleavage. The fact that there is one carbon atom difference between **B** and **C** indicates that **B** is a terminal alkene. Additionally, because the reaction with $KMnO_4$ does not yield an aldehyde, compound **C**—with a degree of unsaturation of one—must be a ketone. With only three carbon atoms available, the only candidate is propanone. Once I establish that compound **C** is propanone, it follows that compound **B** is 2-methylpropene. Furthermore, given that compound **A** is an alcohol that remains unoxidized by $K_2Cr_2O_7$, it must be a tertiary alcohol. Therefore, the addition of H and OH across the C=C bond in compound **B** must occur in such a way as to yield a tertiary alcohol, leading to the formation of compound **A**.

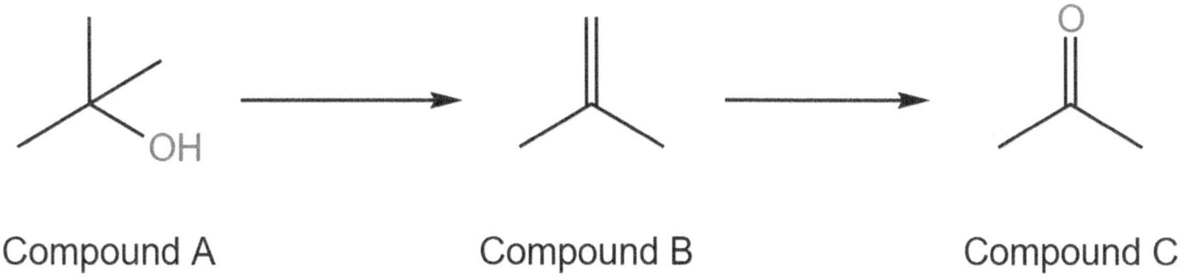

Question 10:

When treated with sodium dichromate(VI), compound **A** ($C_4H_{10}O$) forms a product that does not react with 2,4-dinitrophenylhydrazine but produces effervescence upon the addition of potassium carbonate.

When heated with concentrated sulfuric acid, compound **A** forms compound **B** (C_4H_8), which decolorizes liquid bromine.

Upon reaction with hot acidified potassium permanganate(VII), compound **B** is converted into compound **C** (C_3H_6O), with effervescence observed, and the resulting gas produces a white precipitate when bubbled through calcium hydroxide solution.

Deduce the structures of compounds **A**, **B**, and **C**, and explain carefully what each piece of evidence suggests about their identities.

My thought process:

My reasoning starts by examining compound **A**, which has no degrees of unsaturation and a single oxygen atom. This suggests that **A** is a saturated alcohol. When **A** is oxidized with $Na_2Cr_2O_7$, the resulting product does not react with 2,4-dinitrophenylhydrazine but does produce effervescence with potassium carbonate—behavior that indicates the formation of a carboxylic acid. Since oxidation of a primary alcohol yields a carboxylic acid, this evidence points to **A** being a primary alcohol.

Moving on to compound **B**, we know it is an alkene (C_4H_8). The fact that there is a one carbon atom difference between **B** and compound **C** (C_3H_6O) suggests that **B** is a terminal alkene. When **B** reacts with hot acidified $KMnO_4$, it produces compound **C**, which has one degree of unsaturation and one oxygen atom. Because $KMnO_4$ under these conditions does not form an aldehyde, compound **C** must be a ketone. Given that **C** is a ketone with three carbon atoms, the only possibility is that it is propanone.

Finally, since **A** is derived from **B** by the addition of hydrogen and a hydroxyl group—and it must yield a primary alcohol—the only structure consistent with this pathway is that **A** is 2-methylpropan-1-ol, and **B** is 2-methylpropene, whose oxidative cleavage produces propanone as compound **C**.

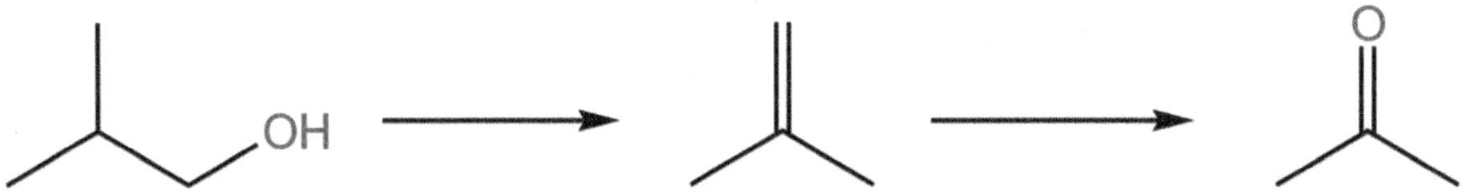

Compound A Compound B Compound C

Question 11:

Compound **A** ($C_4H_{10}O$) reacts with potassium dichromate(VI) to form a product that does not produce an orange precipitate when treated with 2,4-dinitrophenylhydrazine.

When heated with concentrated sulfuric acid, compound **A** forms compound **B** (C_4H_8) only, which decolorizes liquid bromine.

Upon reaction with hot acidified potassium permanganate(VII), compound **B** is converted into compound **C** ($C_3H_6O_2$), with the evolution of a gas.

Deduce the structures of compounds **A**, **B**, and **C**, and explain carefully what each piece of evidence suggests about their identities.

My thought process:

My reasoning begins with compound **A** ($C_4H_{10}O$), which shows no degrees of unsaturation and contains a single oxygen atom, indicating that it is a saturated alcohol. Since **A** reacts with $K_2Cr_2O_7$, it must be either a primary or secondary alcohol. However, because the oxidized product does not yield an orange precipitate when treated with 2,4-dinitrophenylhydrazine, the oxidation must have produced a carboxylic acid rather than an aldehyde or ketone. This result is consistent only with the oxidation of a primary alcohol, leading me to conclude that **A** is a primary alcohol.

Next, compound **B** is formed when **A** is heated with concentrated sulfuric acid; **B** has the formula C_4H_8 and decolorizes liquid bromine, confirming its identity as an alkene. The fact that there is a one-carbon difference between **B** and compound **C** ($C_3H_6O_2$) suggests that **B** is a terminal alkene. The features of compound **C**—a degree of unsaturation of one and two oxygen atoms—indicate that it is a carboxylic acid. Given that **C** contains three carbon atoms, the only possibility is that **C** is propanoic acid. Since propanoic acid is obtained by the oxidative cleavage of a terminal alkene, this confirms that **B** is but-1-ene.

Finally, because compound **A** is produced by the addition of hydrogen and a hydroxyl group across the C=C bond of **B**, and we have already determined that **A** must be a primary alcohol, the addition must occur in a manner that yields butan-1-ol.

Thus, the evidence supports the conclusion that compound **A** is butan-1-ol, compound **B** is but-1-ene, and compound **C** is propanoic acid.

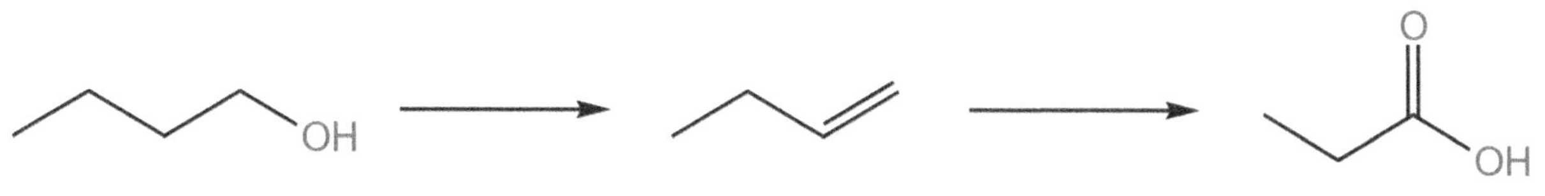

Compound A Compound B Compound C

Question 12:

Upon heating with potassium dichromate(VI) for a prolonged period, compound **A** is oxidized to compound **E**, which gives an orange precipitate with 2,4-dinitrophenylhydrazine.

When heated with concentrated sulfuric acid, compound **A** produces three isomeric alkenes, **B**, **C**, and **D**, all having the molecular formula C_4H_8.

Only compounds **C** and **D**, when treated with hot acidified potassium permanganate(VII), yield the same oxidation product, compound **F** ($C_2H_4O_2$); compound **B** gives organic compound **G**.

Deduce the structures of compounds **A**, **B**, **C**, **D**, **E**, **F**, and **G**, and explain carefully what each piece of evidence suggests about their identities.

My thought process:

My reasoning starts with compound **E**, which gives an orange precipitate when treated with 2,4-dinitrophenylhydrazine. This clearly indicates that **E** contains a carbonyl group—either an aldehyde or a ketone. Since **E** is produced by prolonged oxidation of compound **A** using $K_2Cr_2O_7$, and considering that tertiary alcohols do not oxidize while primary alcohols would typically form carboxylic acids (which would not give this result), it follows that **A** must be a secondary alcohol.

Next, when compound **A** is heated with concentrated sulfuric acid, it undergoes dehydration to yield three isomeric alkenes (**B**, **C**, and **D**), each with the molecular formula C_4H_8. The fact that these alkenes have four carbon atoms indicates that **A** itself is a four-carbon secondary alcohol.

Further evidence comes from the reaction of alkenes **C** and **D** with hot acidified $KMnO_4$, which produces the same oxidation product, compound **F** ($C_2H_4O_2$). The characteristics of **F**—having one degree of unsaturation and containing two oxygen atoms—suggest that it is a carboxylic acid. Since **F** has two carbon atoms, it must be ethanoic acid. The fact that both **C** and **D** yield ethanoic acid upon oxidative cleavage implies that they are cis/trans isomers of but-2-ene.

Although the molecular formula of compound **B** isn't directly specified, the key insight comes from recognizing that **A**, being a four-carbon secondary alcohol, can only be butan-2-ol. This conclusion comes from drawing out all possible isomers for $C_4H_{10}O$ that result in a secondary alcohol—of which butan-2-ol is the sole possibility.

It's important to note that proficiency in drawing skeletal formulas is especially useful for tackling this relatively high-level question, as it allows one to accurately visualize the possible isomers and their interconversions.

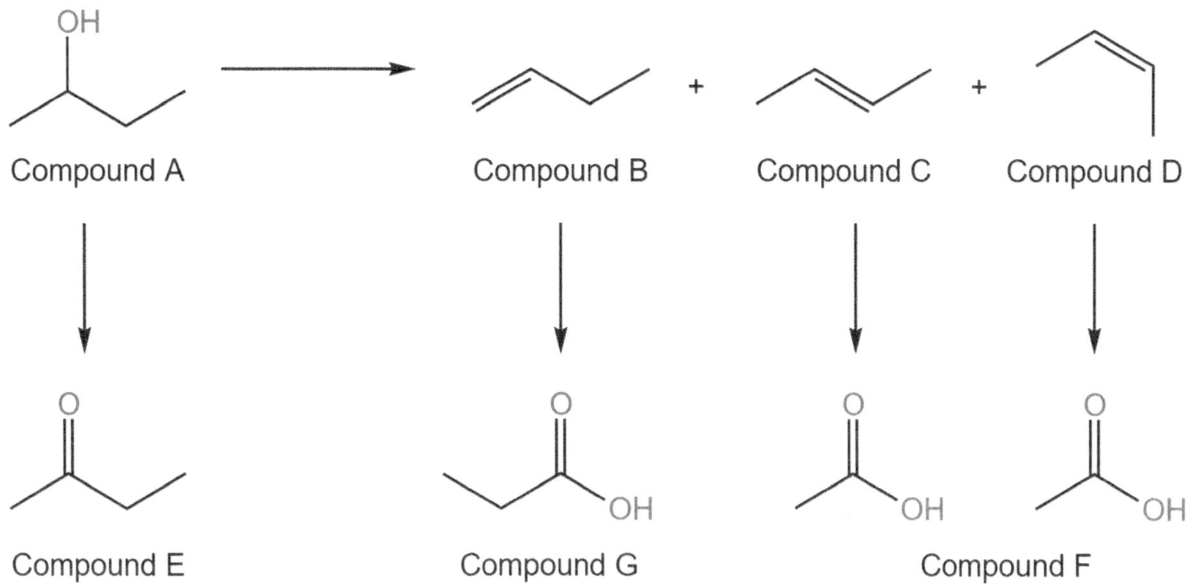
OH
Compound A
+
Compound B
+
Compound C
Compound D
O
Compound E
O
OH
Compound G
O
OH
O
OH
Compound F

Question 13:

Compound **A** ($C_5H_{12}O$) forms a yellow precipitate when treated with alkaline aqueous iodine. When oxidized with potassium dichromate(VI), it forms compound **E**.

When heated with concentrated sulfuric acid, compound **A** forms a pair of geometric isomers, **B** and **C**, and an additional isomer, **D**, all of which decolorize reddish-brown liquid bromine.

Upon reaction with hot acidified potassium permanganate(VII),

- Geometric isomers **B** and **C** yield the same two oxidation products, **F** ($C_3H_6O_2$) and **G** ($C_2H_4O_2$).
- Isomer **D** forms a single organic product, **H** ($C_4H_8O_2$), with the evolution of a gas.

Deduce the structures of compounds **A**, **B**, **C**, **D**, **E**, **F**, **G**, and **H**, and explain carefully what each piece of evidence suggests about their identities.

My thought process:

My analysis begins by considering that compound **A** ($C_5H_{12}O$) has no degrees of unsaturation, a single oxygen atom, and is susceptible to oxidation by $K_2Cr_2O_7$, which indicates that **A** is a saturated alcohol. The fact that **A** gives a positive tri-iodomethane test further confirms it is a secondary alcohol—since among primary alcohols, only ethanol would give this reaction (and ethanol is ruled out because **A** contains five carbon atoms), while tertiary alcohols do not give a positive result either. Consequently, the oxidation of **A** must yield a ketone, which is compound **E**.

When **A** undergoes dehydration, its carbon count remains the same, resulting in the formation of three alkene isomers (**B**, **C**, and **D**), each with five carbon atoms. Notably, the oxidative cleavage of the geometric isomers **B** and **C** produces two oxidation products, **F** and **G**. A key point to remember is that if two alkene isomers give the same oxidation products, they are geometric (cis/trans) isomers. In this case, product **F**, which has one degree of unsaturation and two oxygen atoms, is identified as a three-carbon carboxylic acid (propanoic acid), and product **G**, with two carbon atoms, is recognized as ethanoic acid. Since **F** and **G** together account for all five carbon atoms, this confirms that **B** and **C** are internal alkenes—specifically, pent-2-ene, as only internal alkenes can exhibit cis/trans isomerism.

In contrast, alkene **D** gives a single oxidation product, **H**, along with the evolution of a gas, and **H** has one carbon atom fewer than **D**. This suggests that **D** is a terminal alkene; in this scenario, pent-1-ene, which upon oxidative cleavage yields butanoic acid (**H**).

Taking all the evidence into account, the only structure for **A** that aligns with these observations is pentan-2-ol, a secondary alcohol. Its oxidation forms pentan-2-one (compound **E**); its dehydration produces both pent-1-ene (**D**) and the cis/trans isomers of pent-2-ene (**B** and **C**); and the oxidative cleavage of pent-2-ene results in the formation of propanoic acid (**F**) and ethanoic acid (**G**), while that of pent-1-ene produces butanoic acid (**H**).

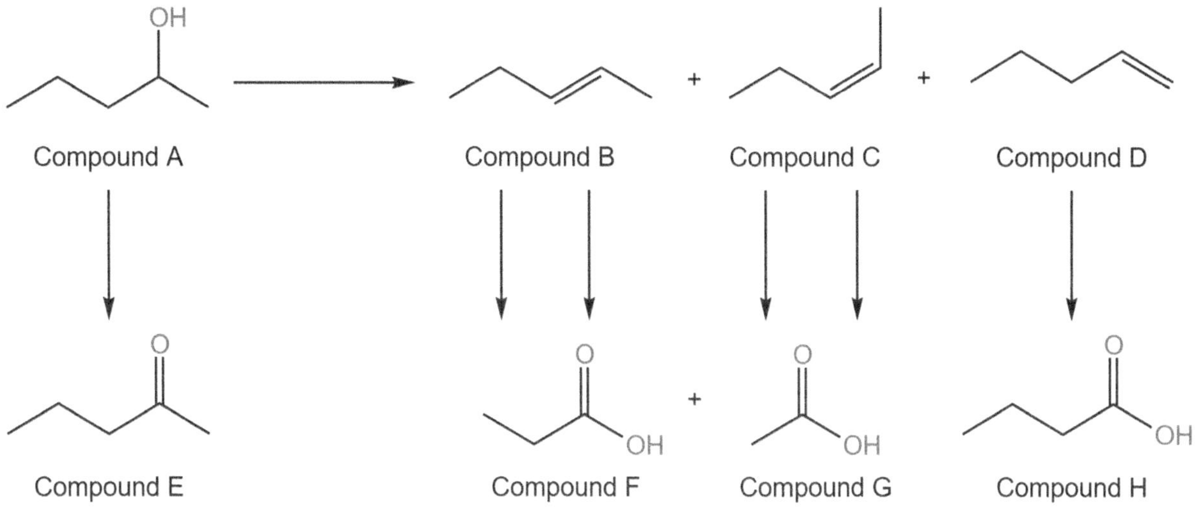

Question 14:

Compound **A** ($C_5H_{12}O$) reacts with potassium dichromate(VI) under reflux to form compound **F** ($C_5H_{10}O$), which produces an orange precipitate when treated with 2,4-dinitrophenylhydrazine.

When heated with concentrated sulfuric acid, compound **A** produces a pair of stereoisomers, compounds **B** and **C**, both of which decolorize liquid bromine.

Upon oxidation, both compound **B** and compound **C** yield the same two organic products, compound **D** ($C_2H_4O_2$) and compound **E** ($C_3H_6O_2$).

Deduce the structures of compounds **A**, **B**, **C**, **D**, **E**, and **F**, and explain carefully what each piece of evidence suggests about their identities.

My thought process:

My analysis begins by noting that compound **A** has no degrees of unsaturation, indicating it is fully saturated. When oxidized with $K_2Cr_2O_7$ under reflux, **A** is converted into a carbonyl compound that gives an orange precipitate with 2,4-DNPH. Since refluxing with $K_2Cr_2O_7$ does not typically produce an aldehyde under these conditions, the oxidation product (**F**) must be a ketone. This observation confirms that **A** is a secondary alcohol.

Next, when **A** undergoes dehydration with concentrated sulfuric acid, its carbon count remains unchanged, yielding five-carbon alkenes. The process produces two stereoisomers, **B** and **C**, which are cis/trans isomers—indicating that they are internal alkenes. Upon oxidative cleavage, both **B** and **C** yield the same two products: **D** and **E**. Here, **D** is identified as a two-carbon carboxylic acid (ethanoic acid) and **E** as a three-carbon carboxylic acid (propanoic acid). Together, these products confirm that **B** and **C** are derived from pent-2-ene.

A key detail arises when considering the addition of hydrogen and hydroxyl groups across the C=C bond in pent-2-ene, which could potentially form either pentan-2-ol or pentan-3-ol—both being secondary alcohols. However, the dehydration of pentan-3-ol would only yield a pair of geometric isomers, whereas pentan-2-ol would produce three alkene isomers. This subtle clue indicates that **A** must be pentan-3-ol.

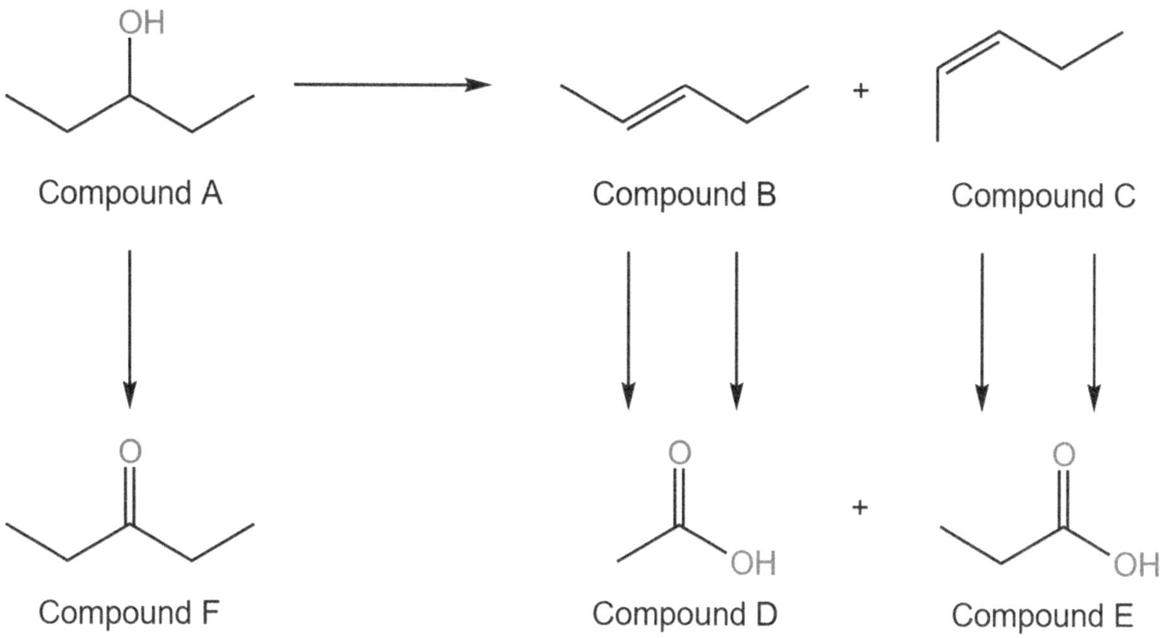
OH
Compound A
+
Compound B
Compound C
O
Compound F
O
OH
Compound D
+
O
OH
Compound E

Question 15:

When treated with potassium dichromate(VI), compound **A** ($C_5H_{12}O$) forms a product that produces an orange precipitate with 2,4-dinitrophenylhydrazine.

Upon reaction with concentrated sulfuric acid, compound **A** forms two isomers, **B** and **C**, both having the molecular formula C_5H_{10}. Both compounds decolorize liquid bromine.

When heated with hot acidified potassium permanganate(VII):

- Compound **B** forms compound **D** ($C_4H_8O_2$), with the evolution of a gas.
- Compound **C** forms two products: compound **E** ($C_2H_4O_2$) and compound **F** (C_3H_6O).

Deduce the structures of compounds **A**, **B**, **C**, **D**, **E**, and **F**, and explain carefully what each piece of evidence suggests about their identities.

My thought process:

My analysis begins by noting that compound **A** has no degrees of unsaturation and contains only one oxygen atom, which tells me it is an alcohol. Since **A** can be oxidized by $K_2Cr_2O_7$, it must be either a primary or a secondary alcohol—primary alcohols would form an aldehyde (or a carboxylic acid under reflux conditions), while secondary alcohols yield ketones. Because $K_2Cr_2O_7$ is a relatively mild oxidant that can only oxidize a primary alcohol to an aldehyde if immediate distillation is performed (otherwise, under reflux, it would convert to a carboxylic acid), and the conditions here aren't specified, I cannot decide immediately whether **A** is primary or secondary. It is crucial not to jump to conclusions at this stage.

When **A** undergoes dehydration with concentrated sulfuric acid, its carbon skeleton remains intact, resulting in the formation of five-carbon alkenes **B** and **C**. The fact that the oxidation products from **B** and **C**, namely compounds **E** and **F**, together account for five carbon atoms helps in constructing the structure of alkene **C**. Here, **E** is identified as a carboxylic acid with two carbon atoms (ethanoic acid), and since oxidative cleavage with $KMnO_4$ will never yield an aldehyde, **F** must be a three-carbon ketone (propanone). This evidence indicates that alkene **C** is 2-methylbut-2-ene. Additionally, there is a four-carbon carboxylic acid, **D**, and the one-carbon difference between alkene **B** and **D** suggests that **B** is a terminal alkene.

Finally, considering the addition of H and OH across the C=C bond in alkene **C** (2-methylbut-2-ene), the only way to obtain an alcohol that, upon dehydration, gives a terminal alkene (like **B**) is if compound **A** is 3-methylbutan-2-ol. Since 3-methylbutan-2-ol is a secondary alcohol, its oxidation with $K_2Cr_2O_7$ will yield a ketone that gives a positive result with 2,4-DNPH.

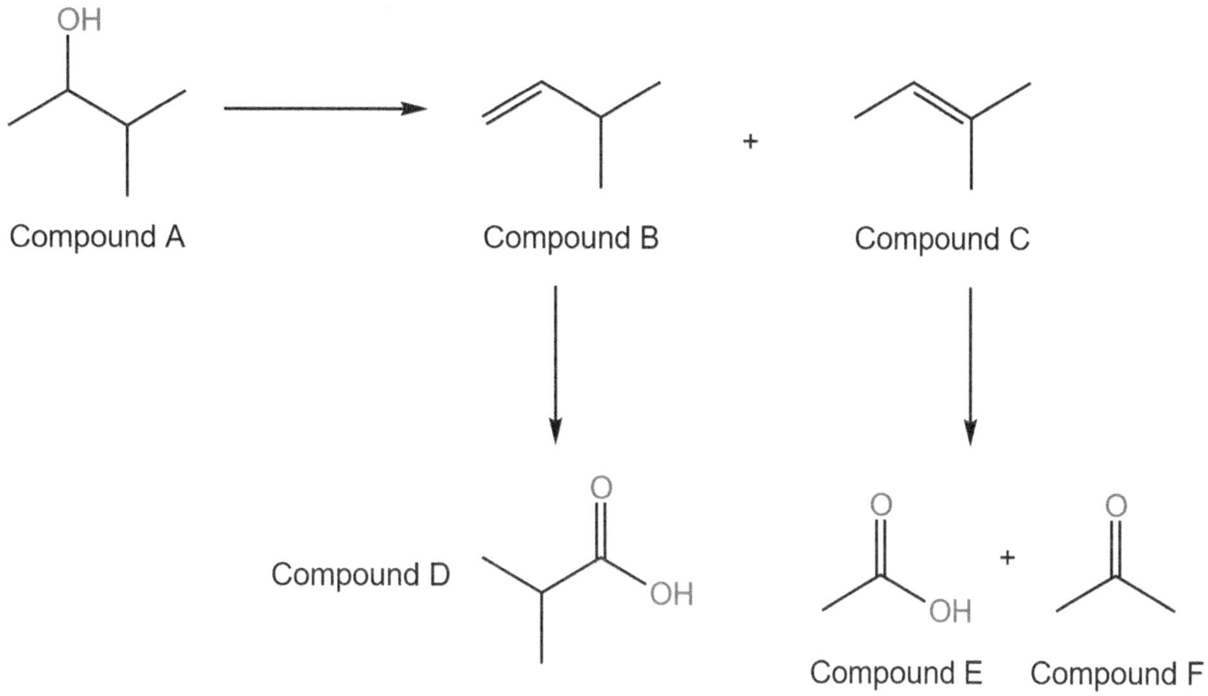
OH
Compound A
Compound B
Compound C
Compound D
O
OH
O
OH
O
Compound E
Compound F

Question 16:

Compound **A** ($C_5H_{12}O$) reacts with potassium dichromate(VI) to form compound **D** ($C_5H_{10}O_2$).

When heated with concentrated sulfuric acid, compound **A** forms compound **B** (C_5H_{10}), which decolorizes liquid bromine.

Upon reaction with acidified potassium permanganate(VII), compound **B** is converted into compound **C** ($C_4H_8O_2$) with the evolution of a gas.

When treated with lithium aluminium hydride, compound **C** forms compound **E** ($C_4H_{10}O$).

Upon heating with hot concentrated sulfuric acid, compound **E** forms compound **F** (C_4H_8).

When treated with hot acidified potassium permanganate(VII), compound **F** is oxidized to compound **G** ($C_3H_6O_2$).

Deduce the structures of compounds **A**, **B**, **C**, **D**, **E**, **F**, and **G**, and explain carefully what each piece of evidence suggests about their identities.

My thought process:

My analysis begins by noting that compound **A** has no degrees of unsaturation and contains a single oxygen atom, so it is a saturated alcohol. When oxidized with $K_2Cr_2O_7$, **A** forms compound **D**, which shows one degree of unsaturation and two oxygen atoms, indicating that **D** is a carboxylic acid. This observation implies that **A** is a primary alcohol.

Dehydrating **A** produces an alkene, compound **B**, while keeping the carbon count unchanged. When **B** is subjected to oxidative cleavage, it forms compound **C**—a carboxylic acid with one degree of unsaturation and two oxygen atoms. The fact that there is a one-carbon difference between **B** and **C** tells me that B is a terminal alkene.

Next, reducing the carboxylic acid **C** with $LiAlH_4$ yields a primary alcohol, compound **E**, which has four carbon atoms. Dehydrating **E** gives an alkene, compound **F**, also containing four carbon atoms. When **F** undergoes oxidative cleavage, it forms compound **G**, which, with one degree of unsaturation and two oxygen atoms, is identified as a carboxylic acid. Since **G** is a three-carbon acid (propanoic acid), and there is a one-carbon difference between **F** (four carbons) and **G**, it confirms that **F** is a terminal alkene.

To piece together the structures, I start with the weakest link: compound **G** is propanoic acid, which must come from the oxidative cleavage of a terminal alkene **F**. Therefore, **F** is but-1-ene. There are two possible ways to add hydrogen and hydroxyl groups across the double bond in but-1-ene to form alcohol **E**; however, since **E** is a primary alcohol, the addition must occur to yield butan-1-ol. If **E** is butan-1-ol, then compound **C**, obtained from the oxidative cleavage of a terminal alkene **B**, is butanoic acid, implying that **B** is pent-1-ene. Since dehydration of primary alcohol **A** yields **B** (pent-1-ene), **A** must be pentan-1-ol, which, upon oxidation, gives pentanoic acid (**D**).

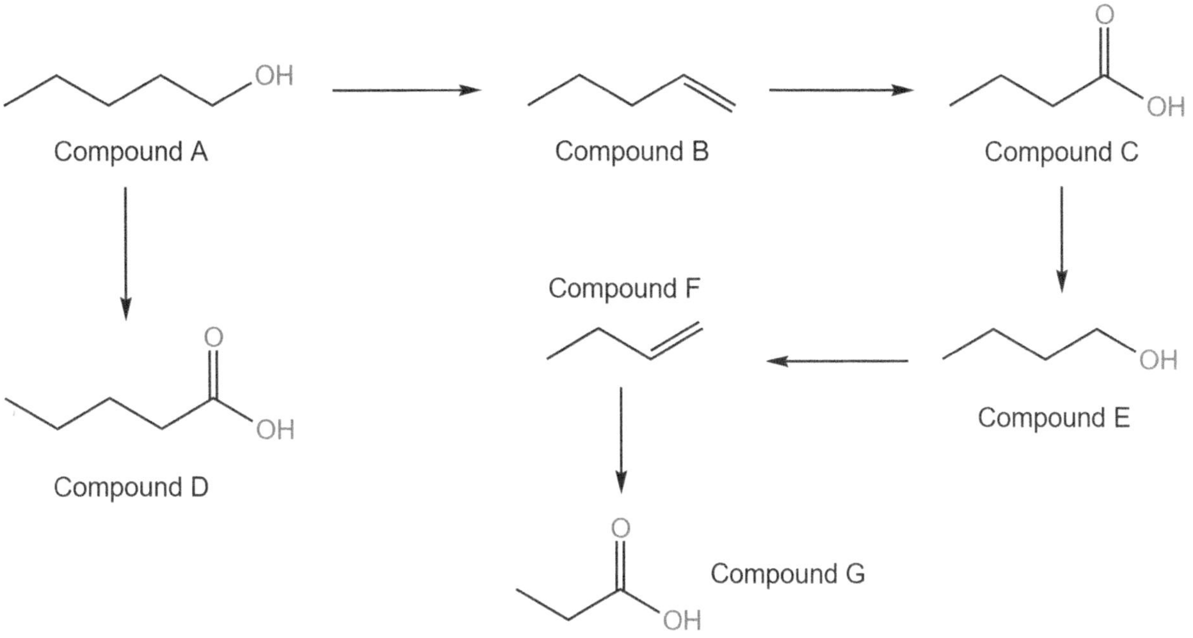

Question 17:

Compound **A** ($C_5H_{12}O$) is converted to compound **B** (C_5H_{10}) when treated with concentrated sulfuric acid. Compound **B** decolorizes liquid bromine.

When treated with potassium permanganate(VII), compound **B** produces a single organic product, **C** (C_4H_8O).

Deduce the structures of compounds **A**, **B**, and **C**, and explain carefully what each piece of evidence suggests about their identities.

My thought process:

My analysis begins by noting that compound **A** has no degrees of unsaturation and contains one oxygen atom, which means it is a saturated alcohol. When **A** undergoes dehydration, it forms an alkene, compound **B**. The subsequent oxidative cleavage of **B** results in a ketone, compound **C**. Since $KMnO_4$ never produces an aldehyde under these conditions, **C** must be a four-carbon ketone—specifically butanone. The fact that there is a one-carbon difference between the ketone and the alkene indicates that **B** is a terminal alkene, namely 2-methylbut-1-ene.

Next, considering the addition of H and OH across the double bond in 2-methylbut-1-ene, there are two possible regioisomeric outcomes. However, since the dehydration of alcohol **A** gives rise to only one alkene product, it suggests that either **A** is symmetrical (so that water removal from either side yields the same alkene) or the hydroxyl group is on the terminal carbon. In this case, the latter is true, meaning that **A** must be 2-methylbutan-1-ol.

🔍 **Tip for students:** When a reaction yields only a single organic product, it greatly limits the possible isomers, making it easier to pinpoint the correct structure—this is the critical clue in this problem.

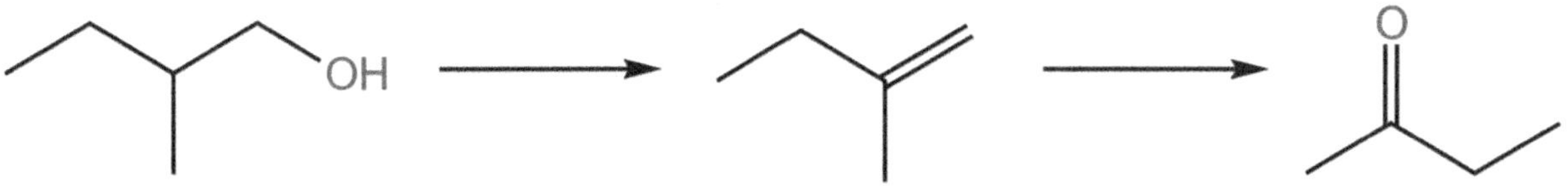

Compound A Compound B Compound C

Question 18:

Compound **A** ($C_5H_{12}O$) does not react with hot concentrated sulfuric acid.

When treated with sodium metal in dry ether, effervescence is observed, and the gas produced extinguishes a lighted splint with a 'pop' sound.

Deduce the structure of compound **A**, and explain carefully what each piece of evidence suggests about its identity.

My thought process:

My reasoning begins by noting that compound **A** has zero degrees of unsaturation and contains one oxygen atom, which suggests it is a saturated alcohol. However, its inability to undergo dehydration indicates that the hydrogen and hydroxyl groups are not located on adjacent carbon atoms. In contrast, when compound **A** is treated with sodium metal, it produces hydrogen gas, confirming the presence of a hydroxyl group. This failure to dehydrate implies that **A** is not a straight-chain alcohol but a branched one. Given that **A** is a five-carbon alcohol, it's relatively simple to draw all possible isomers, and the only structure that fits is 2,2-dimethylpropan-1-ol. This example underscores the importance of being proficient in drawing skeletal formulas when tackling such problems.

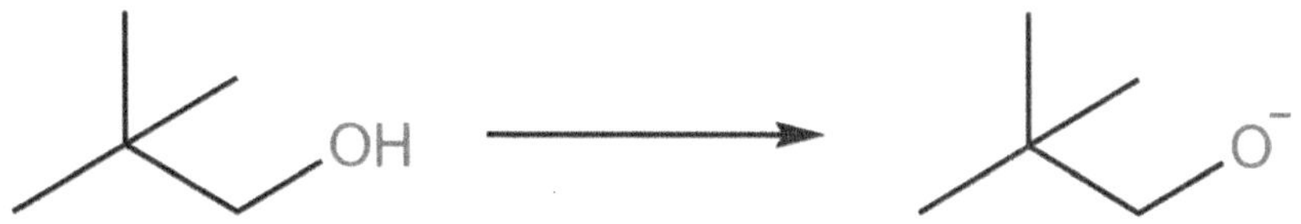

Compound A

Question 19:

Compound **A** ($C_5H_{12}O$) does not react with potassium dichromate(VI).

When heated with hot concentrated sulfuric acid, compound **A** forms two different compounds, **B** and **C**, both having the molecular formula C_5H_{10}.

Upon treatment with hot acidified potassium permanganate(VII):

- Compound **B** forms compound **D** ($C_2H_4O_2$) and compound **E** (C_3H_6O).
- Compound **C** forms only one organic compound, **F** (C_4H_8O).

When treated with 2,4-dinitrophenylhydrazine, only compounds **E** and **F** give a positive test, while compound **D** does not.

Deduce the structures of compounds **A**, **B**, **C**, **D**, **E**, and **F**, and explain carefully what each piece of evidence suggests about their identities.

My thought process:

My reasoning starts by observing that compound **A** shows zero degrees of unsaturation and contains one oxygen atom, indicating it is a saturated alcohol. Its ability to undergo dehydration with hot concentrated sulfuric acid, yielding alkenes **B** and **C**, confirms its alcohol nature. Furthermore, since **A** does not react with $K_2Cr_2O_7$, it is resistant to oxidation under these conditions, which is characteristic of tertiary alcohols. Given that **A** is a five-carbon alcohol and must be tertiary, the only feasible structure is 2-methylbutan-2-ol. Once **A** is identified, determining the structures of compounds **B** through **F** becomes straightforward using standard organic reactions.

Tip for students: When an alcohol does not oxidize or dehydrate to produce multiple isomers, it significantly narrows down the possible structures—this is a key clue for solving the problem.

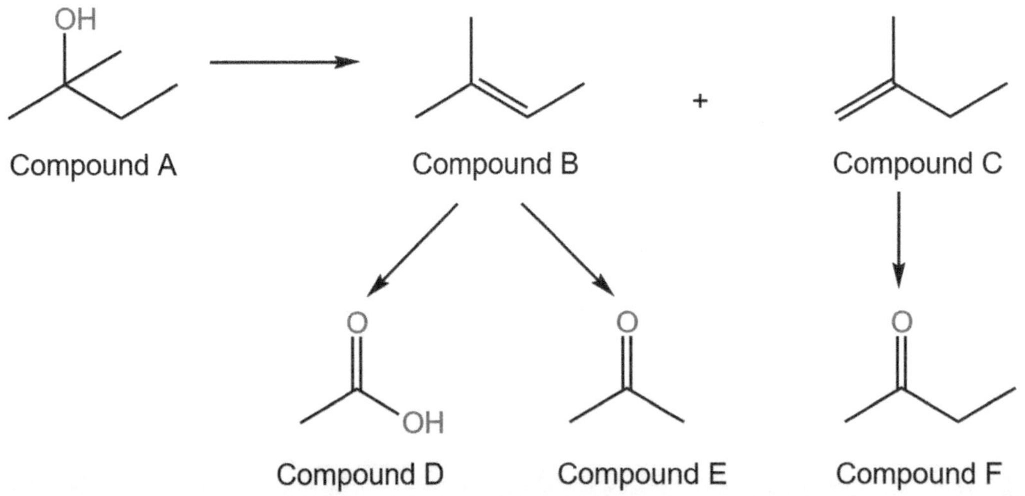

Question 20:

Compound **A** ($C_5H_{12}O$) undergoes treatment with concentrated sulfuric acid, yielding compound **B** (C_5H_{10}), which decolorizes aqueous bromine.

When hot acidified potassium permanganate(VII) is added to compound **B**, it forms compound **C** ($C_4H_8O_2$), which does not form an orange precipitate with 2,4-dinitrophenylhydrazine.

When compound **A** is treated with potassium dichromate(VI), it forms compound **D** ($C_5H_{10}O_2$).

Upon treatment with lithium aluminium hydride, compound **C** is reduced to compound **E** ($C_4H_{10}O$).

When concentrated sulfuric acid is added to compound **E**, it forms compound **F** (C_4H_8).

When treated with hot acidified potassium permanganate (VII), compound **F** produces a single organic compound **G** (C_3H_6O).

Deduce the structures of compounds **A**, **B**, **C**, **D**, **E**, **F**, and **G**, and explain carefully what each piece of evidence suggests about their identities.

My thought process:

My analysis begins by noting that compound **A** has zero degrees of unsaturation and contains one oxygen atom, indicating that it is a saturated alcohol. When **A** is dehydrated, it forms an alkene (compound **B**). The oxidative cleavage of this five-carbon alkene **B** gives rise to a four-carbon carboxylic acid (compound **C**), which suggests that B is a terminal alkene. Additionally, the oxidation of compound **A** with $K_2Cr_2O_7$ produces a carboxylic acid (compound **D**), a transformation typical of primary alcohols; therefore, **A** must be a primary alcohol. Reducing carboxylic acid **C** with $LiAlH_4$ yields a primary alcohol (compound **E**) with four carbon atoms, and subsequent dehydration of **E** forms a four-carbon alkene **F**. When alkene **F** undergoes oxidative cleavage using hot acidified $KMnO_4$, it produces a three-carbon compound **G**. Since strong oxidizing agents like $KMnO_4$ do not generate aldehydes under these conditions, **G** must be a ketone—in this case, propanone.

Working backward from these findings, because the oxidative cleavage of terminal alkene **F** yields propanone, **F** must be 2-methylpropene. The hydration of 2-methylpropene, with appropriate regioselectivity to yield a primary alcohol, leads to compound **E** being 2-methylpropan-1-ol. Moreover, since reducing carboxylic acid **C** gives **E**, compound **C** must be 2-methylpropanoic acid. Similarly, since the oxidative cleavage of terminal alkene **B** results in carboxylic acid **C**, **B** must be 3-methylbut-1-ene. Finally, the hydration of 3-methylbut-1-ene, occurring such that a primary alcohol is formed, confirms that compound **A** is

3-methylbutan-1-ol. Consistently, the oxidation of **A** then produces 3-methylbutanoic acid (compound **D**).

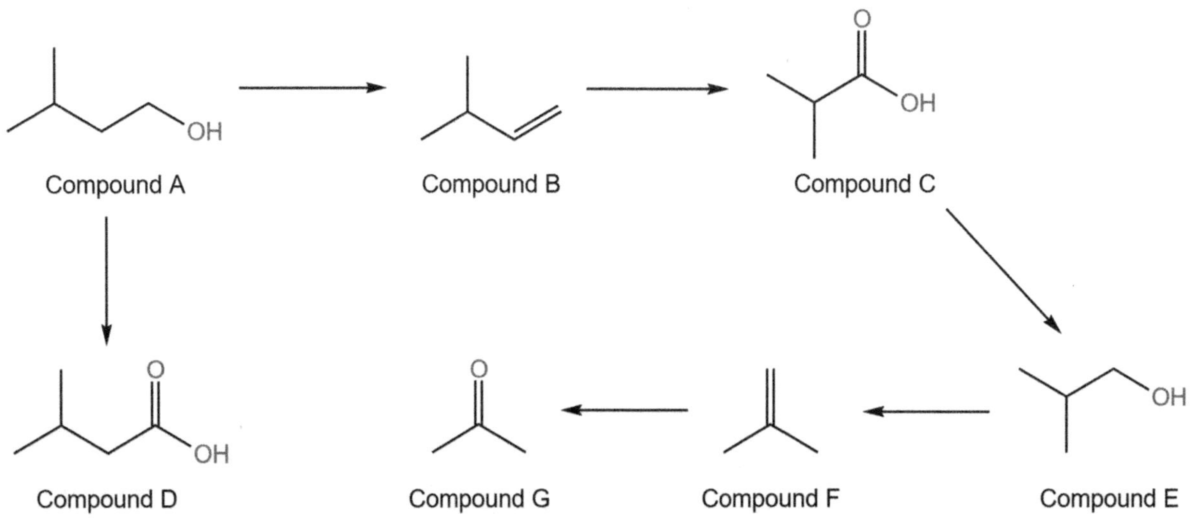

Question 21:

Compound **A** ($C_4H_8O_2$) is heated with sodium hydroxide, producing compound **B** (CH_4O) and an organic salt.

Upon acidic hydrolysis, the organic salt is converted into compound **C**.

When compound **C** is reduced, it forms compound **D**.

Upon heating with concentrated sulfuric acid, compound **D** forms compound **E**.

When compound **E** is treated with hot acidified potassium permanganate(VII), it is converted into compound **F** ($C_2H_4O_2$), which produces effervescence when reacted with sodium hydrogen carbonate.

Deduce the structures of compounds **A**, **B**, **C**, **D**, **E**, and **F**, and explain carefully what each piece of evidence suggests about their identities.

My thought process:

Starting with compound **A**, which has one degree of unsaturation and two oxygen atoms, we initially suspect it might be a carboxylic acid. However, since alkaline hydrolysis of **A** produces two distinct products—a compound **B** and an organic salt—it becomes clear that **A** is actually an ester. The formation of methanol (compound **B**) upon saponification indicates that **A** is methyl propanoate, leaving behind a three-carbon carboxylate ion that, upon acidification, yields compound **C** (propanoic acid).

Next, reducing propanoic acid (**C**) with $LiAlH_4$ converts it into a primary alcohol, identified as propan-1-ol (compound **D**). Dehydration of propan-1-ol using concentrated sulfuric acid then produces an alkene, compound **E** (propene). Finally, when propene (**E**) is treated with hot acidified $KMnO_4$, it undergoes oxidative cleavage to yield compound **F**. Under these strong oxidative conditions—which do not typically produce aldehydes—compound **F** is determined to be ethanoic acid, further confirmed by its effervescence upon reaction with sodium hydrogen carbonate.

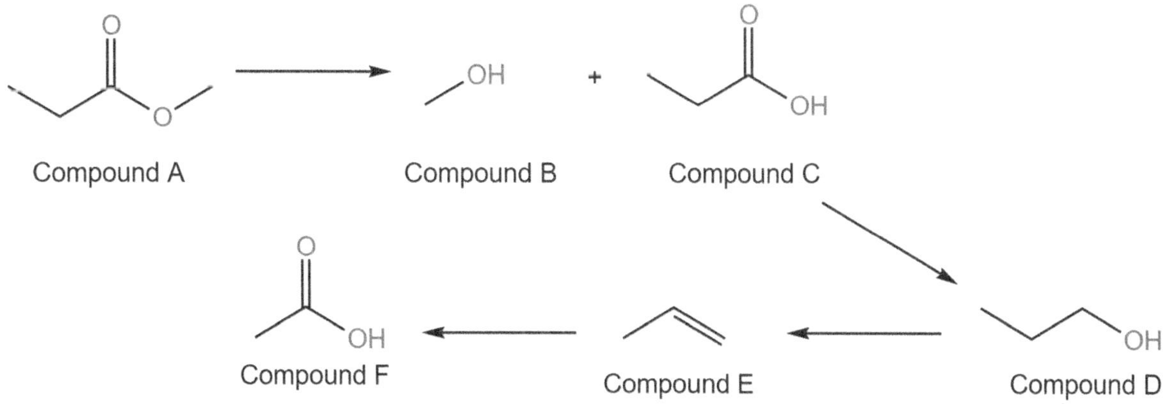

Question 22:

Compound **A** ($C_4H_8O_2$) is heated with sodium hydroxide, followed by acidification, to form compound **B** and compound **C**.

When hot acidified potassium permanganate(VII) is added to compound **B**, only effervescence is observed.

When hot acidified potassium permanganate(VII) is added to compound **C**, it is converted into compound **D**, which does not form an orange precipitate with 2,4-dinitrophenylhydrazine.

Additionally, compound **C** does not form a yellow precipitate when treated with alkaline aqueous iodine.

Deduce the structures of compounds **A**, **B**, **C**, and **D**, and explain carefully what each piece of evidence suggests about their identities.

My thought process:

My analysis begins by noting that compound **A** has one degree of unsaturation and contains two oxygen atoms, which might initially suggest it could be a carboxylic acid. However, because alkaline hydrolysis of **A** produces two distinct products (compounds **B** and **C**), this indicates that **A** is actually an ester. Ester hydrolysis typically yields one carboxylic acid and one alcohol. At this point, it isn't immediately clear which hydrolysis product is the acid and which is the alcohol, since either methanol or methanoic acid could be formed—and both can evolve gas (effervescence) when treated with hot acidified $KMnO_4$.

The crucial clue comes from the reactivity of compound **C**: when **C** is oxidized with $KMnO_4$, it undergoes oxidation, which tells me that **C** is not already a carboxylic acid. Furthermore, compound **D**—formed from oxidizing **C**—fails to give an orange precipitate with 2,4-dinitrophenylhydrazine, confirming that **D** does not contain a reactive carbonyl group such as in aldehydes or ketones but is instead a fully oxidized carboxylic acid. This evidence forces the assignment that compound **C** must be a primary alcohol, and consequently, compound **B** must be the carboxylic acid. Considering that the only carboxylic acid capable of producing effervescence (due to CO_2 evolution) under these oxidation conditions is methanoic acid, compound **B** is identified as methanoic acid.

Given that the total carbon count in compound **A** is four and one carbon is in methanoic acid (**B**), compound **C** must be a three-carbon alcohol; therefore, **C** is propan-1-ol. Confirming this further, propan-1-ol does not produce a yellow precipitate when treated with alkaline aqueous iodine. Finally, oxidizing propan-1-ol (**C**) yields a carboxylic acid, and since propan-1-ol contains three carbons, its oxidation product, compound **D**, is propanoic acid.

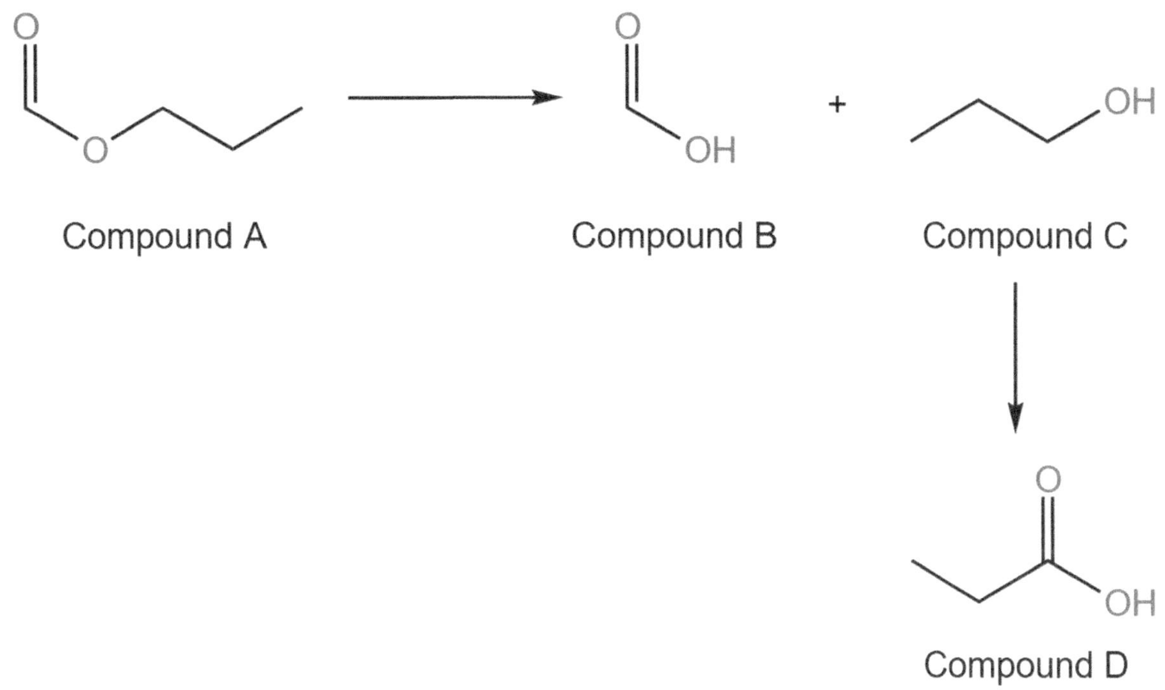
O
O
O
OH
+
OH
Compound A
Compound B
Compound C
O
OH
Compound D

Question 23:

Compound **A** ($C_4H_8O_2$) is heated with aqueous sodium hydroxide, forming compound **B** and compound **C**.

When hot acidified potassium permanganate(VII) is added to compound **B**, only effervescence is observed.

Compound **C** gives a yellow precipitate when treated with hot alkaline iodine.

Deduce the structures of compounds **A**, **B**, and **C**, and explain carefully what each piece of evidence suggests about their identities.

My thought process:

My reasoning begins by noting that compound **A** has one degree of unsaturation and contains two oxygen atoms, which might initially suggest a carboxylic acid. However, since alkaline hydrolysis of **A** yields two distinct products (compounds **B** and **C**), it is clear that **A** is actually an ester. Ester hydrolysis generates both a carboxylic acid and an alcohol. At this stage, it isn't immediately obvious which of the two products is the acid and which is the alcohol—after all, if we had methanol or methanoic acid, both could potentially give effervescence when subjected to hot acidified $KMnO_4$.

The decisive clue comes from the behavior of compound **C**: it gives a yellow precipitate with hot alkaline iodine, a positive iodoform test. This result indicates that C is an alcohol possessing the characteristic $–CH(OH)CH_3$ group, which is typical of a secondary alcohol like propan-2-ol. Since **C** is confirmed as the alcohol, compound **B** must be the carboxylic acid. Given that the only carboxylic acid known to evolve gas (via CO_2 release) under these oxidation conditions is methanoic acid, **B** must be methanoic acid.

Therefore, the original ester **A** is comprised of methanoic acid and propan-2-ol.

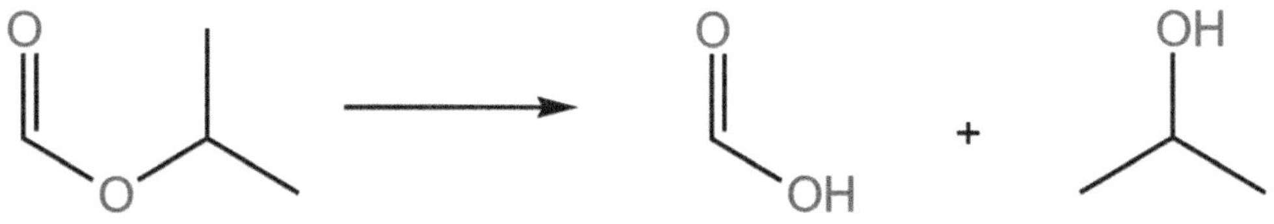

Compound A Compound B Compound C

Question 24:

Compound **A** ($C_4H_8O_2$) is heated with aqueous sodium hydroxide, forming compound **B** and compound **C**.

Compound **C** gives a yellow precipitate when treated with hot alkaline iodine.

When compound **C** is treated with concentrated sulfuric acid, it forms an organic compound that produces only effervescence upon addition of hot acidified potassium permanganate(VII).

Deduce the structures of compounds **A**, **B**, and **C**, and explain carefully what each piece of evidence suggests about their identities.

My thought process:

My analysis begins with compound **A**, which has one degree of unsaturation and contains two oxygen atoms. Although these features might initially suggest a carboxylic acid, the fact that alkaline hydrolysis of **A** yields two distinct products indicates that **A** is actually an ester. Ester hydrolysis produces both a carboxylic acid and an alcohol. At this point, it is not immediately clear which of the resulting compounds is the acid and which is the alcohol.

A critical clue is provided by the behavior of compound **C**: when treated with hot alkaline iodine, **C** forms a yellow precipitate, demonstrating a positive iodoform test. This test is characteristic of alcohols containing the –CH(OH)CH_3 group; for primary alcohols, ethanol is the only one that gives a positive iodoform reaction. Additionally, when compound **C** is treated with concentrated sulfuric acid, it undergoes dehydration to yield an organic compound that, upon oxidation with hot acidified potassium permanganate, produces only effervescence. This outcome is consistent with the dehydration of ethanol to form ethene, which is then oxidized to release carbon dioxide (observed as effervescence).

Therefore, compound **C** is identified as ethanol. Consequently, compound **B**, the other product of the ester hydrolysis, must be the carboxylic acid. Given that the overall carbon count in **A** is four and ethanol contains two carbon atoms, **B** is deduced to be ethanoic acid.

Hence, compound **A** is the ester formed from ethanoic acid and ethanol, which is ethyl ethanoate.

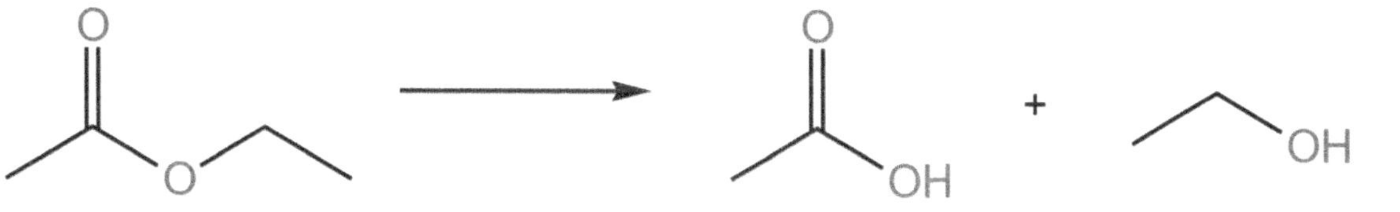

Compound A Compound B Compound C

Question 25:

A liquid compound **A** is oily and insoluble in water. However, when heated with aqueous sodium hydroxide, it dissolves. From the reaction mixture, a new liquid **B** can be obtained by distillation.

Liquid **B** reacts with iodine in sodium hydroxide solution, forming a yellow precipitate. When **B** undergoes controlled oxidation, it is converted into compound **C**, which also gives a yellow precipitate with iodine in sodium hydroxide. Additionally, **C** reacts with 2,4-dinitrophenylhydrazine, forming an orange precipitate.

When sulfuric acid is added to the reaction mixture obtained from heating **A** with sodium hydroxide, a white precipitate **D** is formed. **D** reacts with sodium hydrogen carbonate, releasing one mole of carbon dioxide per mole of **D**, indicating a specific functional group.

When **D** is strongly heated with soda-lime, a new compound is formed, which is identified as benzene. Soda-lime is a mixture of sodium hydroxide and calcium oxide that facilitates the removal of a certain functional group through decarboxylation.

Deduce the structures of **A**, **B**, **C**, and **D**, justifying each step logically based on the given experimental observations. Explain how each piece of evidence supports your reasoning.

My thought process:

I began by observing that compound **A** is an oily liquid and insoluble in water, yet it dissolves when heated with aqueous sodium hydroxide. This change strongly indicates that A undergoes hydrolysis in basic conditions. Since esters are typically nonpolar (and hence oily and water-insoluble) but are converted into water-soluble species through alkaline hydrolysis, this led me to suspect that **A** is an ester. During the hydrolysis, the ester splits into two parts: an alcohol and a carboxylate ion. The alcohol is then isolated by distillation as compound **B**, a process which implies that **B** must be volatile and therefore have a relatively low boiling point; these characteristics are consistent with small molecules like alcohols or aldehydes.

Next, the fact that **B** gives a positive iodoform (triiodomethane) test—evidenced by the formation of a yellow precipitate in the presence of iodine and sodium hydroxide—suggests that **B** contains either a methyl ketone group ($-COCH_3$) or an oxidizable alcohol that could be transformed into such a group. At this stage, **B** could conceivably be an alcohol with a $-CH(OH)CH_3$ moiety or already a carbonyl compound. However, when **B** is subjected to controlled oxidation, the product formed (compound **C**) still gives a positive iodoform test; additionally, **C** reacts with 2,4-dinitrophenylhydrazine to form an orange precipitate, confirming the presence of a carbonyl group. Controlled oxidation of a primary alcohol such as ethanol typically gives ethanal—the only aldehyde that is known to give a positive iodoform test.

Therefore, if oxidation turns **B** into ethanal (compound **C**), then **B** itself must logically be ethanol.

Meanwhile, the hydrolysis of **A** also produces a carboxylate ion, which, upon acidification with sulfuric acid, precipitates as a white solid, compound **D**. The identity of **D** is further confirmed by its behavior: it reacts with sodium hydrogen carbonate to release one mole of carbon dioxide per mole, indicating the presence of a single carboxyl group. Most convincingly, upon decarboxylation with soda lime—a process that removes the carboxyl group—**D** yields benzene. This decarboxylation reaction is a well-known transformation for benzoic acid, meaning that **D** must be benzoic acid.

Putting all these findings together, the only consistent set of structures is as follows: Compound **A** is an ester formed by the condensation of benzoic acid and ethanol, namely ethyl benzoate; its alkaline hydrolysis gives ethanol (**B**) and benzoate ion, which, after acidification, forms benzoic acid (**D**). Ethanol (**B**), when oxidized under controlled conditions, produces ethanal (**C**), which shows the expected carbonyl reactivity.

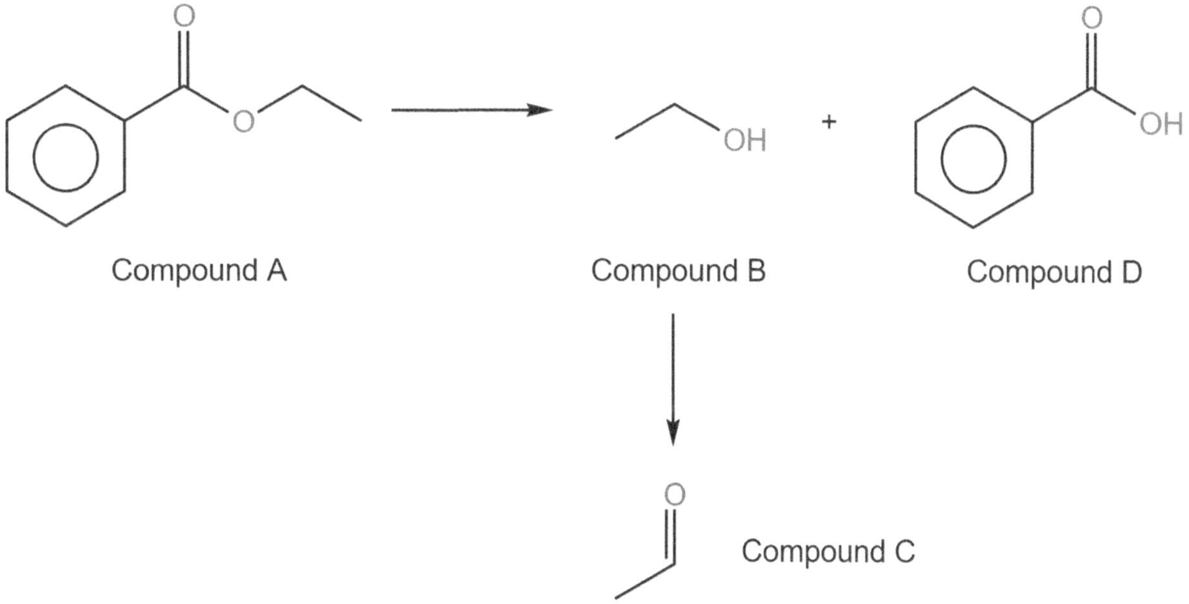

Question 26:

A water-insoluble oily liquid, **A**, has the molecular formula $C_{10}H_{12}O_2$. When heated with aqueous sodium hydroxide, it dissolves. From the resulting mixture, a volatile liquid **B** can be isolated through distillation.

Liquid **B** reacts with iodine in sodium hydroxide solution, forming a yellow precipitate. When subjected to vigorous oxidation, **B** is transformed into compound **C**, which also produces a yellow precipitate under the same iodine-sodium hydroxide test.

Separately, when sulfuric acid is added to the alkaline reaction mixture of **A** and sodium hydroxide, a white precipitate (**D**) forms. **D** reacts with sodium hydrogen carbonate, releasing carbon dioxide in a 1:1 molar ratio, which provides insight into its molecular structure.

Upon strong heating with soda-lime, **D** is converted into benzene, indicating that a specific functional group has been removed. Soda-lime, a combination of calcium oxide and sodium hydroxide, promotes a process called decarboxylation. This reaction eliminates a carboxyl (-COOH) group and replaces it with a hydrogen atom (-H), reducing the molecular complexity of the compound.

Using the experimental evidence provided, determine the structures of **A**, **B**, **C**, and **D**, and explain in detail how each step supports your deductions.

My thought process:

I began by examining compound **A**, a water-insoluble oily liquid with the molecular formula $C_{10}H_{12}O_2$, and calculated its degree of unsaturation to be 5. Since a benzene ring accounts for 4 degrees of unsaturation and an additional degree is typically due to a carbonyl group, this initially pointed toward an aromatic system containing a C=O bond.

When **A** is heated with aqueous sodium hydroxide, it dissolves, indicating that it undergoes alkaline hydrolysis. In an ester hydrolysis, the ester bond is cleaved to yield a carboxylate ion and an alcohol. From the reaction mixture, a volatile liquid **B** is isolated by distillation—a clue that **B** is a small molecule with a low boiling point, though its exact identity is not immediately evident.

Turning my attention to **B**, I note that it gives a positive iodoform test (producing a yellow precipitate when reacted with iodine in sodium hydroxide). This test is characteristic of compounds that possess either a methyl ketone group or a secondary alcohol with a –$CH(OH)CH_3$ fragment, which can be oxidized to a methyl ketone. Indeed, upon vigorous oxidation, **B** is converted into compound **C**, which also gives a positive iodoform test. These observations suggest that **B** is a secondary alcohol that oxidizes to form a methyl ketone.

At this point, the identity of **B** is still ambiguous in terms of its carbon count; it could be any small secondary alcohol. To resolve this, I examine the fate of the other product from **A**'s hydrolysis. When the reaction mixture is acidified with sulfuric acid, a white precipitate **D** forms. This compound reacts with sodium hydrogen carbonate, releasing carbon dioxide in a 1:1 molar ratio, which confirms the presence of a single carboxyl group. Furthermore, when **D** is heated with soda-lime, it undergoes decarboxylation to yield benzene—a reaction characteristic of benzoic acid. This evidence establishes that **D** is benzoic acid, which contains 7 carbon atoms.

Knowing that **A** contains a total of 10 carbon atoms and that the acid fragment (benzoic acid) provides 7 carbons, the remaining fragment—now recognized as the alcohol **B**—must account for the remaining 3 carbons. Given that **B** gives a positive iodoform test and is converted upon oxidation to a methyl ketone (compound **C**), the only three-carbon secondary alcohol fitting these criteria is propan-2-ol, which oxidizes to yield propanone.

Thus, piecing all the evidence together, I conclude that compound **A** is the ester formed by the condensation of benzoic acid and propan-2-ol. Its alkaline hydrolysis produces propan-2-ol (compound **B**) and the benzoate ion, which is then protonated to form benzoic acid (compound **D**). The oxidation of propan-2-ol yields propanone (compound **C**), in agreement with the positive iodoform tests observed.

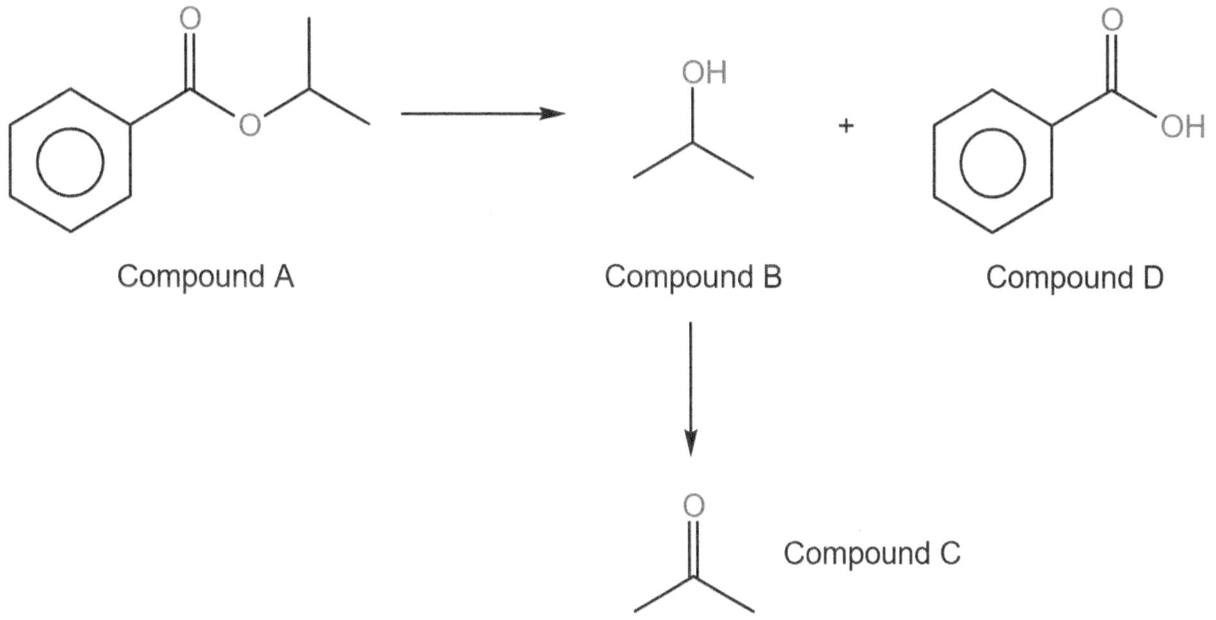

Question 27:

A water-insoluble, oily liquid **A** has the molecular formula $C_{10}H_{12}O_2$. When heated with aqueous sodium hydroxide, it dissolves, forming a new mixture. From this solution, a volatile liquid **B** can be extracted through distillation.

Unlike some compounds with similar properties, **B** does not produce a yellow precipitate when treated with iodine in sodium hydroxide solution.

Separately, when sulfuric acid is added to the alkaline reaction mixture of **A** and sodium hydroxide, a white precipitate (**C**) forms. This solid reacts with sodium hydrogen carbonate, releasing carbon dioxide in a precise 1:1 molar ratio.

Upon strong heating with soda-lime, **C** undergoes a chemical transformation, yielding benzene. This indicates the removal of a particular functional group through decarboxylation, a process in which a carboxyl (-COOH) group is eliminated and replaced by a hydrogen atom (-H). Soda-lime, a mixture of calcium oxide and sodium hydroxide, facilitates this reaction by creating the necessary conditions for the loss of carbon dioxide (CO_2) and the simplification of the molecular structure.

Analyze the given experimental evidence and deduce the structures of **A**, **B**, and **C**. Provide a detailed explanation of how each observation leads to the correct structural identification.

My thought process:

Compound **A** exhibits five degrees of unsaturation, which likely indicates the presence of a phenyl group. Since liquid **B** is volatile and isolated by distillation, it suggests that **B** has a low boiling point—pointing to it being either an alcohol or an aldehyde.

Upon acidification of the hydrolyzed mixture with H_2SO_4, compound **C** is produced, and its reaction with sodium hydrogen carbonate indicates that **C** is a carboxylic acid. Specifically, since one mole of **C** reacts with $NaHCO_3$ to yield one mole of CO_2, this confirms that **C** contains only a single carboxyl group. The fact that decarboxylation of **C** results in benzene confirms that **C** must be benzoic acid.

If **C** is a carboxylic acid, then **B** must be an alcohol for condensation to yield ester **A**. Given that benzoic acid has seven carbon atoms, the alcohol fragment **B** must contain three carbon atoms. Moreover, the absence of a yellow precipitate in the triiodomethane test for **B**—along with its three-carbon composition—strongly suggests that **B** is propan-1-ol. Consequently, ester **A** is formed via the condensation of benzoic acid with propan-1-ol.

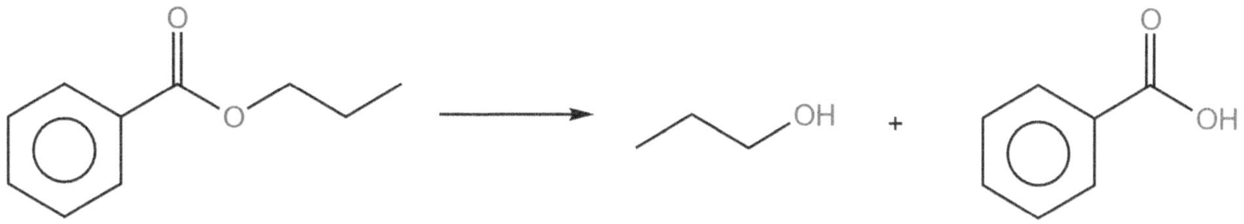

Compound A Compound B Compound C

Question 28:

Compound **A** has a molecular formula of C_4H_8O.

- When sodium metal is added to **A**, effervescence is observed, and the gas produced extinguishes a lighted splint with a 'pop' sound.
- **A** decolorizes aqueous bromine.
- Under reflux, when compound **A** is treated with potassium dichromate(VI), it reacts to produce compound **B** (C_4H_6O), which then gives a positive result when tested with 2,4-dinitrophenylhydrazine.
- When potassium permanganate(VII) is added to **A**, it produces compound **C** ($C_3H_4O_3$) with effervescence.
- **C** forms an orange precipitate with 2,4-dinitrophenylhydrazine.

Deduce the structures of compounds **A**, **B**, and **C**, and explain carefully what each piece of evidence suggests about their identities.

My thought process:

Compound **A** has one degree of unsaturation and contains one oxygen atom. Its reaction with sodium metal, which evolves hydrogen gas, confirms the presence of a hydroxyl group. Additionally, the fact that **A** decolorizes aqueous bromine verifies the presence of an unsaturation; therefore, **A** is an unsaturated alcohol.

Under reflux, **A** is converted into compound **B**, which gives a positive result with 2,4-dinitrophenylhydrazine—this indicates that **B** contains a carbonyl group. However, because reflux conditions would oxidize an aldehyde further to a carboxylic acid, **B** must be a ketone. This, in turn, means that **A** was originally an unsaturated secondary alcohol.

When **A** is treated with $KMnO_4$, the reagent serves two roles: it oxidizes the secondary alcohol to the corresponding ketone and also performs oxidative cleavage of the carbon–carbon double bond. Since there is a loss of one carbon atom upon this oxidation, it implies that **A** is a terminal alkene alcohol. Given that **A** is a four-carbon molecule, a terminal alkene, and a secondary alcohol, these clues together allow us to identify **A**. Once A is established, the identities of **B** and **C** follow naturally.

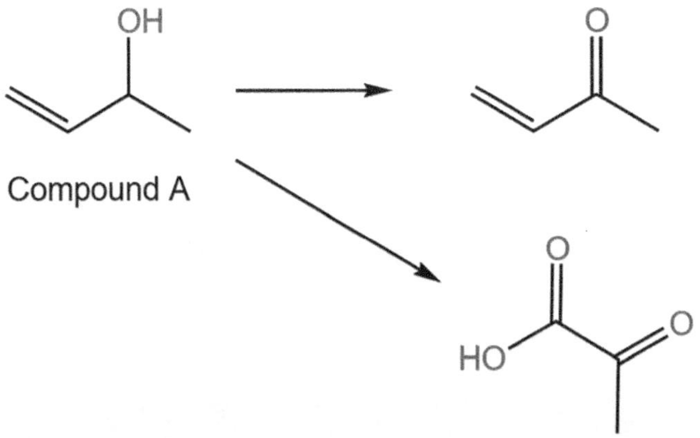

Compound A

Compound B

Compound C

Question 29:

Compound **A** has a molecular formula of C_4H_8O.

- When sodium metal is added to **A**, effervescence is observed, and the gas produced extinguishes a lighted splint with a 'pop' sound.
- **A** decolorizes aqueous bromine.
- Under reflux, when compound **A** is treated with potassium dichromate(VI), it reacts to produce compound **B** ($C_4H_6O_2$).
- When potassium permanganate(VII) is added to **A**, it produces compound **C** ($C_3H_4O_3$) with effervescence.
- **C** forms an orange precipitate with 2,4-dinitrophenylhydrazine.

Deduce the structures of compounds **A**, **B**, and **C**, and explain carefully what each piece of evidence suggests about their identities.

My thought process:

Compound **A** has one degree of unsaturation and contains one oxygen atom. When reacted with sodium metal, it produces hydrogen gas, confirming that it is an alcohol. Additionally, because **A** decolorizes aqueous bromine, we can deduce that it is an unsaturated alcohol.

Compound **B**, on the other hand, exhibits two degrees of unsaturation and contains two oxygen atoms. It is formed under reflux with the oxidizing agent $K_2Cr_2O_7$, which leads us to infer that **B** is an unsaturated carboxylic acid. Given that **B** is a carboxylic acid, **A** must be an unsaturated primary alcohol.

When an unsaturated primary alcohol like **A** is treated with $KMnO_4$, the reaction serves two purposes: first, it cleaves the alkene (C=C) bond oxidatively, and second, it oxidizes the primary alcohol to form a carboxylic acid. Since there is a one-carbon difference between the original alkene **A** and compound **C**, we can conclude that **A** is a terminal alkene primary alcohol.

Compound **C** has two degrees of unsaturation and contains three oxygen atoms, which implies that it possesses both a ketone and a carboxyl group. Considering that **C** is a three-carbon molecule featuring these functionalities, its structure can be deduced accordingly. Once the identity of compound **C** is established, the structures of compounds **A** and **B** follow naturally.

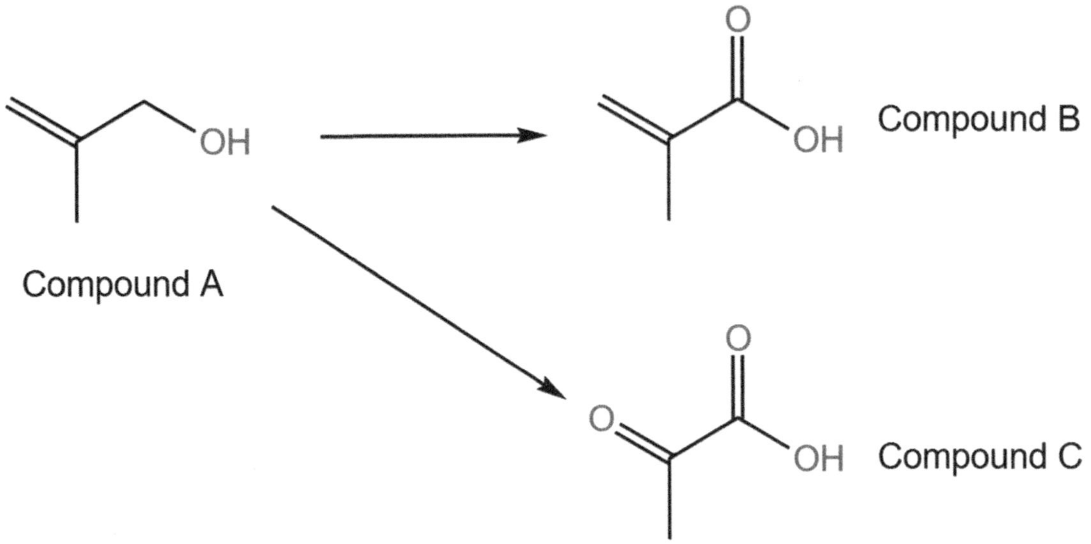

Compound A

Question 30:

Compound **A** has a molecular formula of C_4H_8O.

- When sodium metal is added to **A**, effervescence is observed, and the gas produced extinguishes a lighted splint with a 'pop' sound.
- **A** decolorizes aqueous bromine.
- Under reflux, when compound **A** is treated with potassium dichromate(VI), it reacts to produce compound **B** ($C_4H_6O_2$).
- When potassium permanganate(VII) is added to **A**, it produces compound **C** ($C_2H_4O_2$) with effervescence.

Deduce the structures of compounds **A**, **B**, and **C**, and explain carefully what each piece of evidence suggests about their identities.

My thought process:

Compound **A** has a degree of unsaturation of one and contains a single oxygen atom. Its reaction with sodium metal, which produces hydrogen gas, confirms the presence of a hydroxyl group. Additionally, because **A** decolorizes aqueous bromine, we can deduce that it is an unsaturated alcohol.

Compound **B**, on the other hand, has a degree of unsaturation of two and contains two oxygen atoms. One degree comes from the alkene (C=C) bond—since $K_2Cr_2O_7$ does not oxidize an alkene—while the other arises from the carboxyl group (-COOH). Therefore, **B** is an unsaturated carboxylic acid. This, in turn, implies that **A** must be a primary alcohol because only a primary alcohol is capable of being oxidized into a carboxylic acid. Thus, **A** is an unsaturated primary alcohol.

🔍 **Tip for students:** When a saturated molecule contains two oxygen atoms, there are generally three possibilities: (1) it can be a diol, (2) it might contain two carbonyl groups, or (3) it could be a carboxylic acid (or a derivative such as an ester). Mastering the degree of unsaturation is key—for example, a diol typically shows a DU of 0, a molecule with two carbonyl groups has a DU of 2, and a carboxylic acid exhibits a DU of 1.

Returning to the problem, treatment of **A** with $KMnO_4$ serves two functions: (1) it performs oxidative cleavage of the alkene (C=C) bond, and (2) it oxidizes the primary alcohol group to a carboxylic acid. Compound **C** is produced in this reaction, and it has a degree of unsaturation equal to one and contains two oxygen atoms. This indicates that **C** must be a two-carbon carboxylic acid—in other words, **C** is ethanoic acid.

Since there is a two-carbon difference between ethanoic acid (compound **C**) and the original alkene **A**, **A** cannot be a typical terminal alkene.

🔍 **Tip for students:** If one carbon atom is lost during oxidative cleavage, it's a strong indication that the alkene is terminal. However, if two carbon atoms are lost, consider two possibilities: either the alkene is terminal at both ends of a long chain, or a C_2 fragment such as ethanedioic acid has formed.

In this case, the evidence supports the latter scenario, revealing that the alkene **A** (combined with the primary alcohol functionality) matches the structure anticipated from ethanedioic acid formation.

Since one half of **A** can now be understood from the identity of compound **C**, the full structure of alkene **A** can be deduced by considering the formation of ethanedioic acid. Once **A** is worked out, the structure of **B** follows naturally.

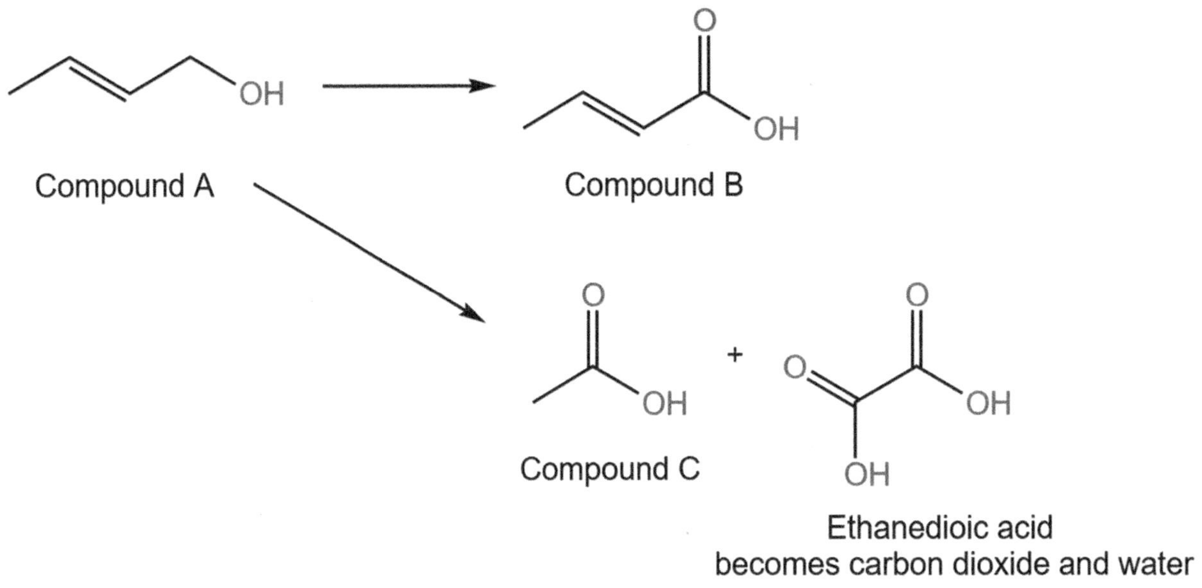

14 Answers to Chapter 11 — C, H, O, and X

The same approach applies here: use the flowcharts to check your proposed structures and challenge yourself to deduce the missing reagents and conditions.

Question 31:

Compound **A** has the chemical formula C_8H_{10}. When exposed to chlorine under ultraviolet light, it reacts to produce compound **B** (C_8H_9Cl). Additionally, **A** undergoes a separate reaction with chlorine under alternative conditions to yield compound **E**, an isomer of **B**.

Compound **B** reacts with aqueous sodium hydroxide to form compound **C** ($C_8H_{10}O$). On the other hand, when **E** is heated with aqueous sodium hydroxide, no reaction takes place. Heating compound **C** with acidified potassium permanganate (VII) results in the formation of compound **D** ($C_7H_6O_2$). Furthermore, reflux of **C** with acidified potassium dichromate (VI) leads to the formation of compound **G**, which reacts with 2,4-dinitrophenylhydrazine to produce an orange precipitate.

When compound **E** is heated with acidified potassium permanganate (VII), effervescence occurs, and compound **F** ($C_7H_5O_2Cl$) is produced. By piecing together this information, deduce the structures of compounds **A** through **G** and explain the logic behind each determination.

My thought process:

The molecular formula of compound **A** indicates a degree of unsaturation of four, which strongly suggests the presence of a benzene ring. The key question becomes how the two remaining carbon atoms are arranged. Potential candidates for **A** include ethylbenzene and the three xylene isomers: 1,2-, 1,3-, and 1,4-dimethylbenzene.

When **A** undergoes chlorination under free radical conditions, it gives compound **B**, an alkyl chloride. Under alternative chlorination conditions, a different product, compound **E**, is obtained. This second route clearly refers to electrophilic aromatic substitution, meaning **E** must be a chlorinated aromatic compound—specifically, a chlorobenzene derivative.

B reacts with aqueous sodium hydroxide, resulting in the formation of an alcohol, **C**, through nucleophilic substitution. However, **E** does not react under the same conditions, which further supports the conclusion that **E** is a chlorobenzene derivative.

The oxidation of compound **A** yields a product with the molecular formula $C_7H_6O_2$, which should immediately suggest benzoic acid to any trained eye. This product confirms that the two carbon atoms beyond the benzene ring are part of a single side chain, most likely an ethyl group. If **A** had been one of the xylene isomers (with two separate methyl groups), oxidation would have produced a compound with the formula $C_8H_6O_4$—a benzene dicarboxylic acid. Thus, we can confidently identify compound **A** as ethylbenzene.

A particularly interesting question arises when considering where the chlorine atom is introduced during the free radical substitution of ethylbenzene. Two possible positions exist: at the primary carbon of the ethyl side chain or at the secondary carbon. This distinction is crucial because it determines whether the final alcohol **C** will be primary or secondary.

The answer lies in a later step: when alcohol **C** is heated under reflux with acidified potassium dichromate(VI), it forms a carbonyl compound. Given that the reaction mixture is refluxed and not distilled immediately, any primary alcohol would be oxidised all the way to a carboxylic acid. However, since the product is a ketone, this tells us that the alcohol must be secondary, and therefore, the chlorine substitution in **B** must have occurred at the secondary carbon of the ethyl group. This is consistent with typical reactivity patterns in free radical halogenation, where more stable secondary radicals are preferentially formed.

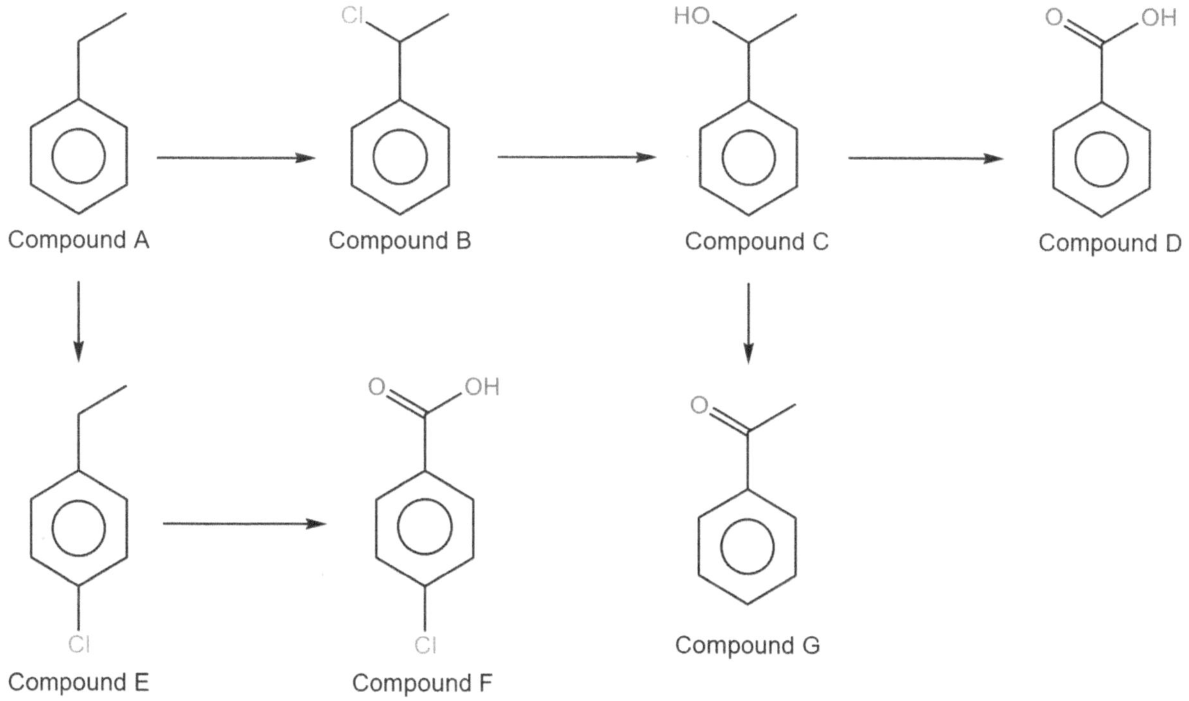

Question 32:

Compound **A** has the molecular formula C_8H_{10}. Upon treatment with chlorine under ultraviolet light, it produces compound **B**, which has the formula C_8H_9Cl. Under different reaction conditions, **A** reacts with chlorine again to form compound **E**, an isomer of **B**.

When compound **B** is reacted with aqueous sodium hydroxide, it yields compound **C**, with the formula $C_8H_{10}O$. However, compound **E** does not undergo any reaction when heated with aqueous sodium hydroxide. Heating compound **C** with acidified potassium permanganate (VII) results in the formation of compound **D** ($C_7H_6O_2$). Additionally, reflux of **C** with acidified potassium dichromate (VI) leads to the production of compound **G** ($C_8H_8O_2$).

When compound **E** is heated with acidified potassium permanganate (VII), effervescence is observed, and compound **F** ($C_7H_5O_2Cl$) is formed. Based on this information, deduce the structural formulas of compounds **A** through **G** while providing an explanation for each deduction.

My thought process:

The molecular formula of compound **A** reveals four degrees of unsaturation, which strongly points to the presence of a benzene ring. The challenge lies in determining how the remaining two carbon atoms are connected to the ring. Possible structures for **A** include ethylbenzene and the three dimethylbenzene isomers—1,2-, 1,3-, and 1,4-dimethylbenzene.

When compound **A** is treated with chlorine under free radical conditions, it forms compound **B**, an alkyl chloride. A separate chlorination using different conditions leads to compound **E**, suggesting an electrophilic substitution reaction has taken place. This clearly indicates that **E** is a chlorinated aromatic compound, more specifically, a chlorobenzene derivative.

Compound **B** undergoes nucleophilic substitution with aqueous sodium hydroxide to form an alcohol, **C**. However, **E** remains unreactive under these same conditions, reinforcing the conclusion that the chlorine atom in **E** is directly attached to the aromatic ring and not to the side chain.

Oxidation of compound **A** yields a compound with the molecular formula $C_7H_6O_2$, which is strongly indicative of benzoic acid. This outcome confirms that the two carbon atoms are part of a single side chain, likely an ethyl group, rather than two separate methyl groups. If **A** had instead been any of the dimethylbenzene isomers, oxidation would have resulted in a benzene dicarboxylic acid ($C_8H_6O_4$). Therefore, we can confidently conclude that compound **A** is ethylbenzene.

A particularly noteworthy aspect is the position at which the chlorine atom is introduced during the free radical chlorination of ethylbenzene. There are two possible sites: the primary carbon (–CH_3) or the secondary carbon (–CH_2–). This distinction determines whether the resulting alcohol **C** will be primary or secondary.

The answer is provided in a subsequent step. When alcohol **C** is refluxed with acidified potassium dichromate(VI), the product is a carboxylic acid, compound **G**. Its molecular formula confirms that it contains two oxygen atoms and has a degree of unsaturation equal to five—four from the aromatic ring and one from the newly formed carboxyl group. This outcome confirms that **C** is a primary alcohol, which could only have arisen from halogenation at the primary carbon of the ethyl group.

At first glance, this may seem to contradict the general expectation that secondary free radicals are more stable than primary radicals, and thus more likely to form. However, in structure elucidation questions, conclusions must be based on the evidence provided in the data—not solely on prior assumptions or trends. While secondary radicals are generally favoured, this does not mean primary radicals cannot be formed under the given conditions.

Finally, compound **E**, the chlorinated aromatic compound, also undergoes side chain oxidation to yield compound **F**, further supporting the structure of **A** as ethylbenzene.

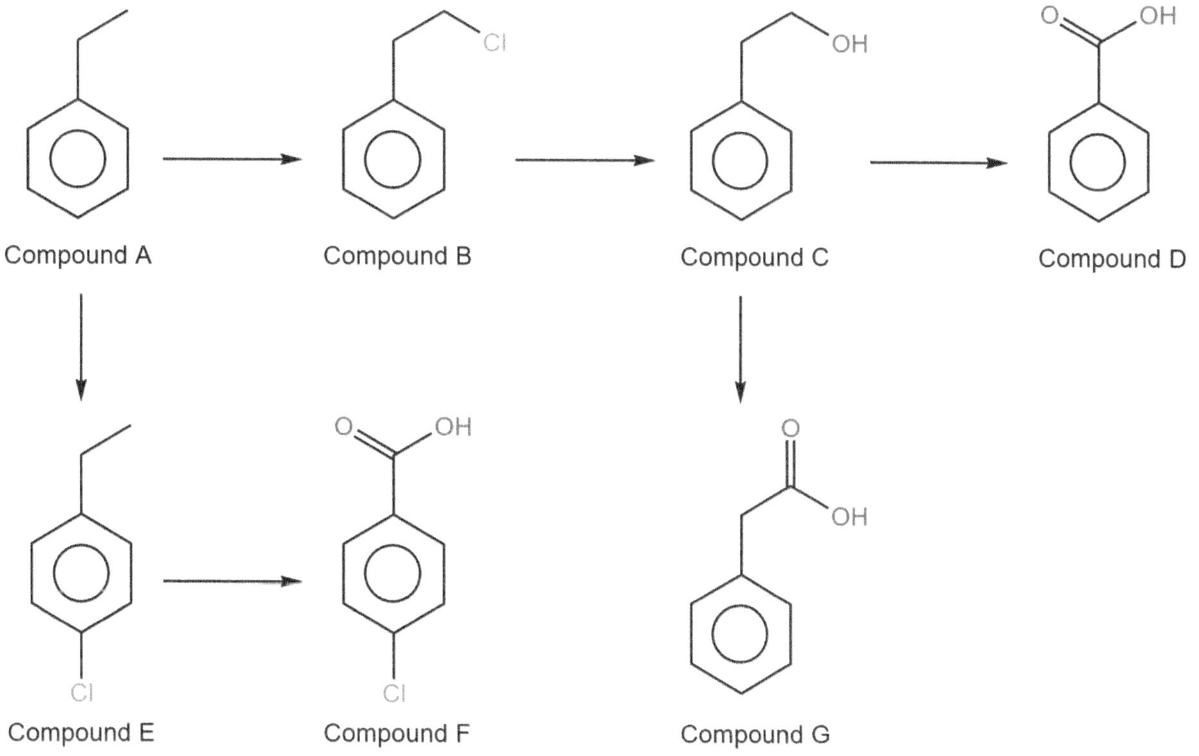

Question 33:

Compound **A** possesses the molecular formula C_8H_{10}. When exposed to chlorine under ultraviolet light, it forms compound **B** with the formula C_8H_9Cl. Additionally, **A** reacts with chlorine under alternate conditions to produce compound **E**, which is isomeric with compound **B**.

Upon treating compound **B** with aqueous sodium hydroxide, compound **C** ($C_8H_{10}O$) is obtained. In contrast, heating compound **E** with aqueous sodium hydroxide results in no reaction. When compound **C** undergoes heating with acidified potassium permanganate (VII), it yields compound **D** ($C_8H_6O_4$). Furthermore, heating compound **C** with acidified potassium dichromate (VI) leads to the formation of compound **G** ($C_8H_8O_2$), and upon strong oxidation, compound **G** is converted back into compound **D**.

When compound **E** is treated with acidified potassium permanganate (VII), decolorization is observed, leading to the formation of compound **F** ($C_8H_5O_4Cl$). Using this information, deduce the structural formulas of compounds **A** through **G** and provide a detailed explanation for each step.

My thought process:

The molecular formula of compound **A** indicates four degrees of unsaturation, strongly suggesting the presence of a benzene ring. The next step is to determine how the remaining two carbon atoms are arranged. Based on this, plausible structures for **A** include ethylbenzene and the three dimethylbenzene isomers: 1,2-, 1,3-, and 1,4-dimethylbenzene.

When **A** is chlorinated under free radical conditions, it produces compound **B**, an alkyl chloride. However, chlorination under alternative conditions yields compound **E**, which clearly results from electrophilic aromatic substitution. This points to **E** being a chlorobenzene derivative, with the chlorine atom directly bonded to the aromatic ring.

Compound **B** reacts with aqueous sodium hydroxide to give alcohol **C** via nucleophilic substitution, whereas compound **E** does not react under the same conditions. This further confirms that **E** is an aryl halide, resistant to nucleophilic substitution, and supports the structural difference between **B** and **E**.

Another key observation comes from compound **D**, whose molecular formula is $C_8H_6O_4$. This is immediately recognisable as a benzene dicarboxylic acid, indicating the presence of two carboxyl groups on a benzene ring. The degree of unsaturation here is six: four from the aromatic ring and two from the carboxyl groups. This clearly shows that the original two carbon atoms must have been present as two distinct methyl groups, each oxidised into a –COOH group, rather than part of a single side chain. Hence, compound **A** must be a dimethylbenzene isomer.

At this point, it remains unclear which of the three isomers—1,2-, 1,3-, or 1,4-dimethylbenzene—compound **A** is. Since no further data is provided to distinguish between them, any of the three isomers would be acceptable at this stage.

From here, compound **C** can be identified as a benzene derivative containing both a primary alcohol and a methyl side chain. When **C** is subjected to mild oxidation using potassium dichromate(VI), only the primary alcohol is converted into a carboxylic acid, while the methyl group remains intact. This gives compound **G**, which contains one –COOH group and one methyl substituent. A subsequent strong oxidation with hot potassium manganate(VII) will then oxidise the remaining methyl group into another carboxyl group, completing the transformation into the benzene dicarboxylic acid structure seen in compound **D**.

Since the question does not provide information that distinguishes which specific dimethylbenzene isomer **A** must be, it is reasonable to accept any one of the three. Later substitutions on the aromatic ring can be predicted based on the known directing effects of substituents, unless the question provides specific information to suggest otherwise.

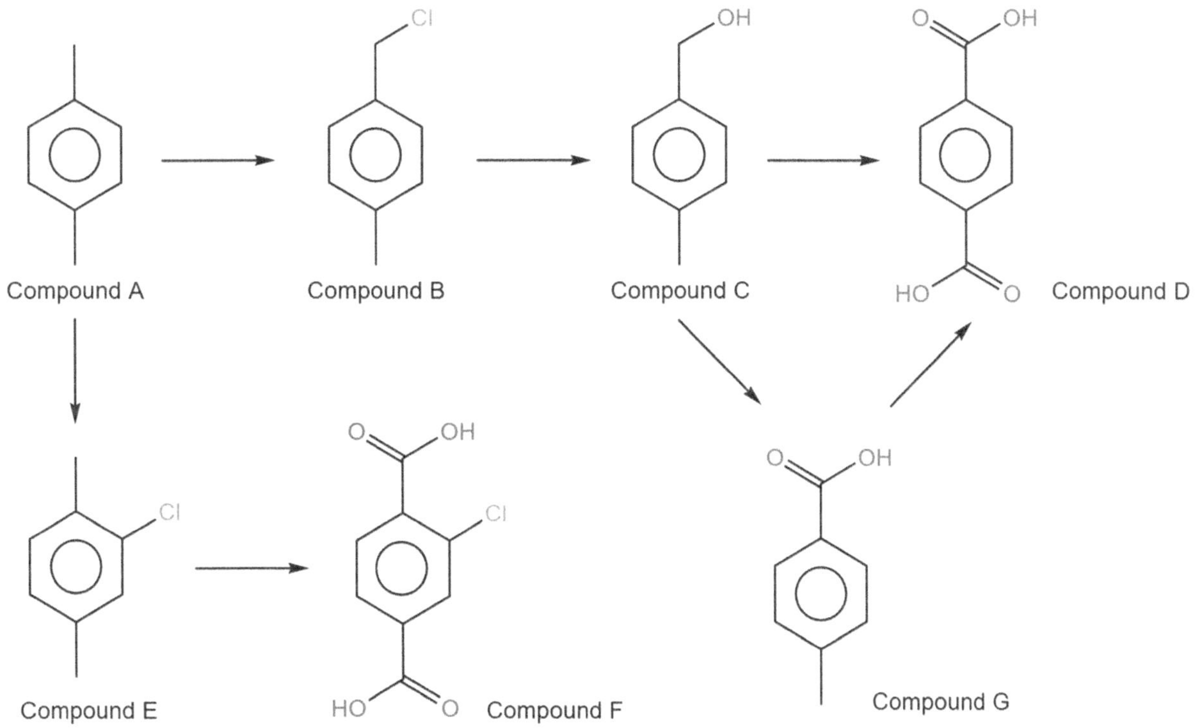

Question 34:

Compound **A**, with the formula $C_4H_8O_2$, reacts with 2,4-dinitrophenylhydrazine to yield an orange precipitate and forms a silver mirror when treated with Tollen's reagent. When treated with acid, compound **A** converts into compound **B** (C_4H_6O), which can exist as either cis or trans isomers. Compound **B** reacts with hydrogen chloride to produce compound **C** (C_4H_7OCl) and undergoes reaction with chlorine to form compound **D** ($C_4H_6OCl_2$).

Upon reduction, compound **C** is transformed into compound **E** (C_4H_9OCl), which, when hydrolyzed using dilute alkali, results in compound **F** ($C_4H_{10}O_2$). Both compounds **A** and **F** give a positive result in the iodoform test (CHI_3), while compound **E** does not react positively. Determine the identities and structures of compounds **A** through **F**, and justify your reasoning.

My thought process:

Compound **A** has a degree of unsaturation equal to one and contains two oxygen atoms. From this alone, one might reasonably suspect that **A** contains both a hydroxyl group and a carbonyl group. This initial deduction highlights the usefulness of analysing degrees of unsaturation when narrowing down structural possibilities.

A gives positive results with both 2,4-dinitrophenylhydrazine and Tollens' reagent, confirming the presence of an aldehyde group. Combined with the earlier assumption that **A** also contains an alcohol group, it's likely that compound **A** is a hydroxyaldehyde.

When **A** is treated with acid, it is converted into compound **B**, which has a degree of unsaturation equal to two—one more than **A**. This additional unsaturation likely arises from the formation of a double bond via acid-catalysed dehydration of the alcohol group in **A**, suggesting that **B** is an alkene. The use of concentrated sulfuric acid in excess typically promotes this transformation.

Compounds **C** and **D** are then formed from **B** via reaction with hydrogen chloride and chlorine gas, respectively—classic electrophilic addition reactions of alkenes. Compound **C**, in particular, undergoes reduction, converting the aldehyde functional group into a primary alcohol, resulting in compound **E**. **E** then reacts with aqueous alkali, leading to a nucleophilic substitution in which the chlorine atom is replaced by a hydroxyl group, forming compound **F**—a diol with a degree of unsaturation equal to zero.

At this stage, we have a good understanding of the transformations and the functional groups involved in compounds **A** through **F**. However, identifying their precise structures remains a challenge—until we reach the critical clue: both **A** and **F** give positive results with the iodoform test. This test is highly specific for the presence of a $–CH(OH)CH_3$ or $–COCH_3$ moiety.

Since **A** is known to contain an alcohol group and gives a positive result with the iodoform test, it must contain the $-CH(OH)CH_3$ group. Given that **A** contains only four carbon atoms, this information leaves us with just one possible structure. Once the structure of **A** is confirmed, the exact position at which dehydration occurs may still seem ambiguous. However, a key clue lies in the fact that compound **B** exists as a pair of geometric isomers (cis and trans). This implies that the double bond in **B** must be formed between two distinct carbon atoms capable of supporting such isomerism—meaning that only one position in **A** can undergo dehydration to give rise to this situation. Conveniently, this same elimination also leads to the formation of an α,β-unsaturated aldehyde, fully consistent with the identity of compound **B**.

The next subtle point arises in the electrophilic addition of HCl to **B**, forming compound **C**. The C=C double bond in **B** presents two possible sites for protonation, and students should be cautious not to jump to conclusions based solely on Markovnikov's rule. In structure deduction questions, early assumptions can mislead. Instead, focus on the experimental evidence: compound **F**, the final diol, also gives a positive iodoform test, which narrows down the possible arrangements of substituents in intermediate compounds.

This final clue confirms the correct structure of compound **C**, and from there, the structures of compounds **E** and **F** follow naturally.

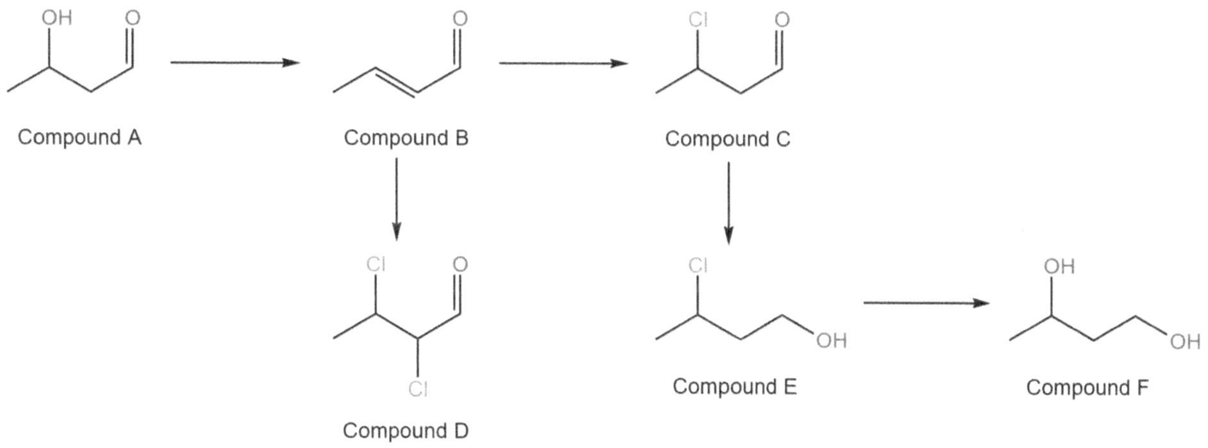

Question 35:

Compound **A**, with the molecular formula $C_4H_8O_2$, reacts with 2,4-dinitrophenylhydrazine to produce an orange precipitate and forms a silver mirror when treated with Tollen's reagent. On treatment with acid, compound **A** is converted to compound **B** (C_4H_6O), which exists as cis and trans isomers. Compound **B** reacts with hydrogen chloride to form compound **C** (C_4H_7OCl) and reacts with chlorine to give compound **D** ($C_4H_6OCl_2$).

Upon reduction, compound **C** is transformed into compound **E** (C_4H_9OCl), which is hydrolyzed using dilute alkali to yield compound **F** ($C_4H_{10}O_2$). While compound **A** tests positive in the iodoform reaction (CHI_3), compounds **E** and **F** do not. Determine the structural identities of compounds **A** through **F**, providing a detailed explanation for your reasoning.

My thought process:

Compound **A** contains two oxygen atoms and has one degree of unsaturation. From this information alone, it is reasonable to infer the presence of both a hydroxyl and a carbonyl group—an observation that illustrates how valuable the degree of unsaturation can be when narrowing down structural possibilities.

A gives positive tests with both 2,4-dinitrophenylhydrazine and Tollens' reagent, confirming that it possesses an aldehyde functional group. When combined with the earlier suspicion that **A** also contains a hydroxyl group, the most likely identity of **A** is that of a hydroxyaldehyde.

Treatment of **A** with acid yields compound **B**, which now has two degrees of unsaturation—an increase of one. This additional unsaturation likely arises from the elimination of water through acid-catalysed dehydration of the hydroxyl group, forming a C=C double bond. The use of concentrated sulfuric acid in excess typically facilitates such transformations, suggesting that **B** is an alkene.

When **B** reacts with hydrogen chloride and chlorine gas, it forms compounds **C** and **D**, respectively—standard electrophilic addition reactions typical of alkenes. Compound **C** can then be reduced, converting its aldehyde group into a primary alcohol to give compound **E**. Subsequent reaction of **E** with aqueous alkali leads to the replacement of the chlorine atom by a hydroxyl group, forming compound **F**, a diol with no remaining degrees of unsaturation.

At this point, while we have established a clear picture of the transformations and functional groups present in compounds **A** through **F**, their exact structures still require clarification. The turning point comes with a critical observation: compound **A** gives a positive iodoform test, while compounds **E** and **F** do not. This is significant, as the iodoform test is selective for compounds containing the $-CH(OH)CH_3$ or $-COCH_3$ moiety.

Since **A** contains an alcohol group and yields a positive iodoform test, it must feature the –CH(OH)CH_3 structure. Given that **A** contains only four carbon atoms, only one structure satisfies all the criteria. Once **A** is identified, attention turns to the dehydration process. While it might initially appear that dehydration could occur at any one of the two sites, the fact that compound **B** exists as two geometric isomers (cis and trans) narrows down the possibilities. This indicates that the double bond must be located between two carbons capable of displaying geometric isomerism, meaning only one viable dehydration site exists. This reaction also conveniently generates an α,β-unsaturated aldehyde, consistent with the identity of **B**.

An additional subtlety emerges in the electrophilic addition of HCl to compound **B**, forming compound **C**. The C=C double bond in **B** presents two different sites for protonation, which can lead to different products. Students are cautioned not to assume that Markovnikov's rule automatically dictates the outcome in structure determination problems. Premature assumptions can lead to incorrect conclusions. Instead, one should rely on experimental data: compound **F** gives a negative iodoform test, which helps confirm the arrangement of substituents throughout the pathway.

It's worth noting that compound **E**, a four-carbon alcohol, does not give a positive result with the iodoform test. This is expected, as **E** is a primary alcohol, and ethanol is the only primary alcohol known to give a positive iodoform test. Therefore, attention should be focused on compound **F**, which does not yield a positive test.

This final clue confirms the identity of compound **C**, allowing the structures of compounds **E** and **F** to be deduced with confidence.

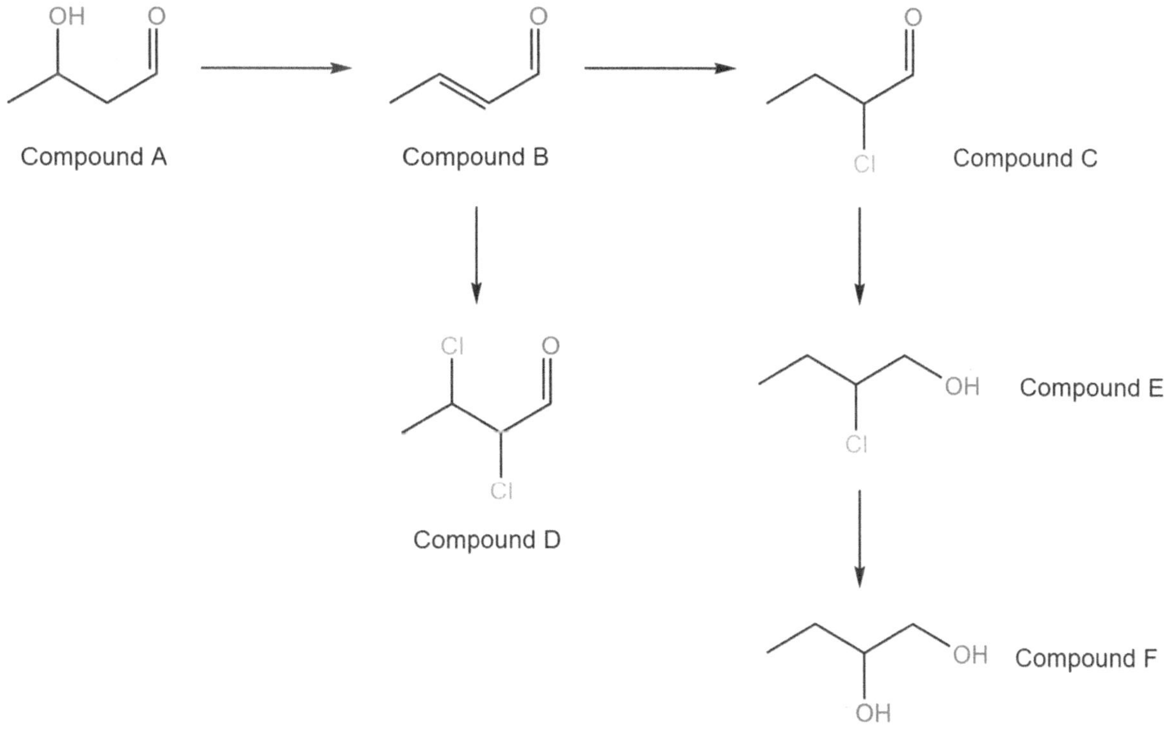

Question 36:

Compound **A** ($C_4H_8O_2$) produces an orange precipitate when treated with 2,4-dinitrophenylhydrazine and forms a silver mirror upon reaction with Tollen's reagent. On exposure to acid, **A** is converted into compound **B** (C_4H_6O), which does not exhibit stereoisomerism. Compound **B** reacts with hydrogen chloride to produce compound **C** (C_4H_7OCl) and undergoes chlorination to yield compound **D** ($C_4H_6OCl_2$).

Reduction of compound **C** results in the formation of compound **E** (C_4H_9OCl), which is hydrolyzed using dilute alkali to give compound **F** ($C_4H_{10}O_2$). Compound **F** gives a positive result in the iodoform reaction (CHI_3), while compounds **A** and **E** do not. Using this information, deduce the structures of compounds **A** through **F** and provide a detailed explanation of the reasoning behind each step.

My thought process:

Compound **A** contains two oxygen atoms and exhibits one degree of unsaturation. From this alone, it is reasonable to propose the presence of both a hydroxyl group and a carbonyl group, highlighting how degree of unsaturation can guide structural deductions.

A gives positive results with 2,4-dinitrophenylhydrazine and Tollens' reagent, confirming the presence of an aldehyde. When considered alongside the earlier suspicion of a hydroxyl group, it strongly suggests that compound **A** is a hydroxyaldehyde.

Upon treatment with acid, **A** is converted into compound **B**, which now possesses two degrees of unsaturation—one more than **A**. This new unsaturation likely results from acid-catalysed dehydration, where the hydroxyl group is eliminated to form a C=C double bond. This is consistent with the use of excess concentrated sulfuric acid, a common dehydrating agent. Interestingly, **B** does not exhibit cis–trans isomerism, which points to a structure where the double bond is likely formed at a terminal position. Given that **B** contains only four carbon atoms, this observation suggests that the -OH group in **A** was on a primary carbon, where geometric isomerism is not possible.

When **B** undergoes electrophilic addition with hydrogen chloride and chlorine, it forms compounds **C** and **D**, respectively—both products of standard electrophilic addition to alkenes. Compound **C** can then be reduced, converting its aldehyde group into a primary alcohol, resulting in compound **E**. Treating **E** with aqueous alkali leads to nucleophilic substitution, replacing the chlorine atom with a hydroxyl group and forming compound **F**—a diol with zero degrees of unsaturation.

Further insight into the structure of these compounds is provided by the iodoform (triiodomethane) test. Compound **A** does not give a positive result, supporting the earlier conclusion that it contains a primary alcohol, not a –CH(OH)CH_3 group. Similarly, compound **E**, which is formed by reduction of the aldehyde group in **C**, is also a primary alcohol and, as expected, does not give a positive iodoform test either.

However, compound **F** does yield a positive result with the iodoform test, which confirms the presence of a –CH(OH)CH_3 moiety—a secondary alcohol. Since **F** is also known to contain a primary alcohol (retained from the reduction of the aldehyde) and has a total of four carbon atoms, its structure can be determined with a high degree of certainty.

A key mechanistic point arises in the addition of HCl to compound **B**. The C=C bond presents two possible sites for protonation, each potentially leading to a different product. The final confirmation of the substitution pattern comes from the structure of compound **F**, which, through its secondary alcohol group, helps trace the sequence backward. From the known structure of **F**, the structures of both compound **E** and compound **C**—which must be secondary alkyl halides—can be confidently deduced.

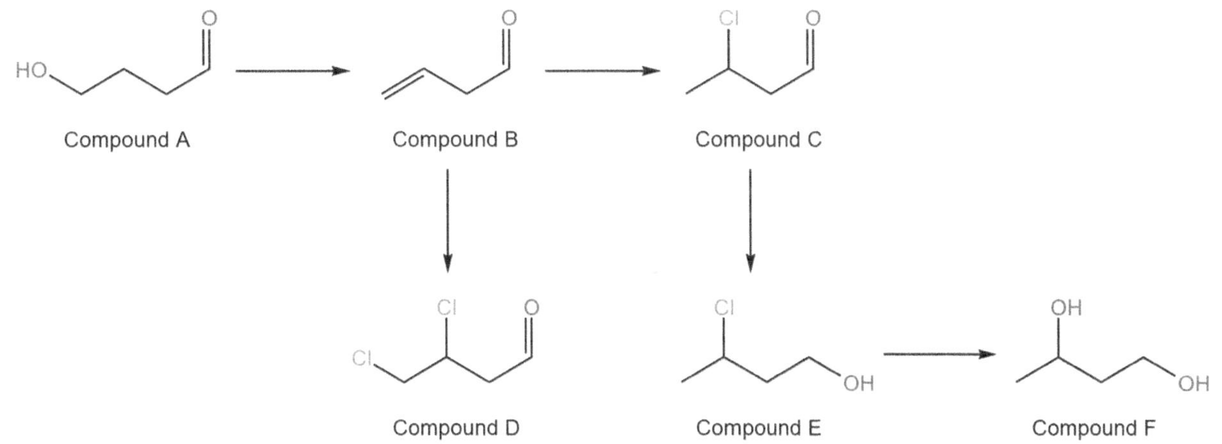

Question 37:

Compound **A** ($C_4H_8O_2$) reacts with 2,4-dinitrophenylhydrazine to form an orange precipitate and produces a silver mirror with Tollen's reagent, indicating the presence of an aldehyde functional group. Upon treatment with acid, **A** transforms into compound **B** (C_4H_6O), which does not exhibit stereoisomerism. Compound **B** reacts with hydrogen chloride to produce compound **C** (C_4H_7OCl) and undergoes chlorination to yield compound **D** ($C_4H_6OCl_2$).

Reduction of compound **C** produces compound **E**, identified as 4-chloro-butan-1-ol, which is then hydrolyzed with dilute alkali to form compound **F** ($C_4H_{10}O_2$). Compounds **A**, **E**, and **F** do not give a positive result in the iodoform test (CHI_3). Using this information, deduce the structures of compounds **A** through **F** and provide a detailed explanation for each step of your reasoning.

My thought process:

Compound **A** contains two oxygen atoms and has one degree of unsaturation. It produces positive results with both 2,4-dinitrophenylhydrazine and Tollen's reagent, indicating the presence of an aldehyde group. Since two oxygen atoms are present, it is likely that **A** also includes a hydroxyl group, making it a hydroxy-aldehyde.

Compound **B**, formed from **A**, has two degrees of unsaturation—an increase of one compared to **A**. This suggests that acid-catalysed dehydration of the hydroxyl group has taken place, most likely under excess concentrated sulfuric acid, resulting in the formation of an alkene. Thus, **B** contains both an alkene and an aldehyde functional group.

Only one alkene product is obtained from this dehydration, and it does not exhibit cis–trans isomerism, indicating that **A** is a primary alcohol. This limits the dehydration pathway, allowing elimination to occur in only one way.

Reactions of **B** with hydrogen chloride and chlorine follow standard electrophilic addition mechanisms, yielding compounds **C** and **D**, respectively. Compound **C** contains both an alkyl chloride and an aldehyde, while **D** is a dichloroalkane that also retains the aldehyde group.

Reduction of the aldehyde group in **C** leads to compound **E**, a primary alcohol that still contains a chlorine atom. The identification of **E** as 4-chlorobutan-1-ol is especially important, as it allows us to deduce the identities of compounds **A** through **D** by reasoning backwards.

Rather than requiring students to determine the structure of **E** through chemical tests, the question provides the IUPAC name directly. This shifts the challenge to interpreting the name correctly and drawing the corresponding structure. In A Level Chemistry, students are expected to recognise and draw structures from IUPAC names, even if naming from a structure is not emphasized to the same degree.

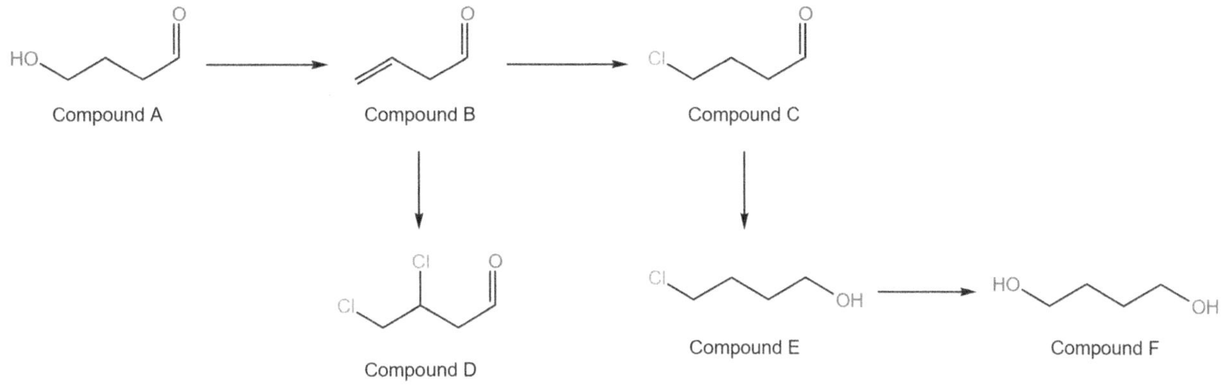

Question 38:

Compound **A** ($C_4H_8O_2$) reacts with 2,4-dinitrophenylhydrazine to form an orange precipitate and produces a silver mirror when treated with Tollen's reagent. When subjected to acidic conditions, **A** is converted into compound **B** (C_4H_6O), which may exist as cis- and trans-isomers. Compound **B** reacts with hydrogen chloride to yield compound **C** (C_4H_7OCl) and undergoes chlorination to produce compound **D** ($C_4H_6OCl_2$).

Reduction of compound **C** results in the formation of compound **E** (C_4H_9OCl), which is hydrolyzed in the presence of dilute alkali to give compound **F** ($C_4H_{10}O_2$). Compounds **A**, **E**, and **F** do not give a positive result in the iodoform test (CHI_3). Strong oxidation of compound **A** produces compound **G**, with the molecular formula $C_4H_6O_3$. Use this information to deduce the structural identities of compounds **A** through **G**, providing clear reasoning for each step in the process.

My thought process:

The fact that compound **A** produces positive results with both 2,4-dinitrophenylhydrazine and Tollens' reagent confirms the presence of an aldehyde functional group. Given that **A** contains two oxygen atoms and has one degree of unsaturation, it is likely a hydroxy aldehyde. For a four-carbon compound with both a hydroxyl and aldehyde group, possible candidates include: 2-hydroxybutanal, 3-hydroxybutanal, 4-hydroxybutanal, 2-hydroxy-2-methylpropanal, and 3-hydroxy-2-methylpropanal.

Upon dehydration, **A** forms compound **B**, which contains both a C=C double bond and an aldehyde. Since **B** displays geometrical (cis-trans) isomerism, it suggests that the hydroxyl group in **A** was not on a primary carbon, as such positions wouldn't yield alkenes capable of exhibiting stereoisomerism. This eliminates any structure where the alcohol is primary. Hence, **A** cannot be 4-hydroxybutanal or 3-hydroxy-2-methylpropanal.

Reactions of **B** with hydrogen chloride and chlorine further confirm the presence of the alkene, as these reagents readily undergo electrophilic addition across C=C bonds. The resulting compound **C** is a saturated halogenoaldehyde. Reduction of **C** transforms the aldehyde into a primary alcohol, producing compound **E**, which still retains the halogen atom. **E** then undergoes alkaline hydrolysis, where the halide is substituted by a hydroxyl group to yield a diol, compound **F**.

A critical point in the identification of **A** is its negative result with the iodoform test. This implies the absence of the –CH(OH)CH_3 group, ruling out structures like 3-hydroxybutanal. Additionally, **A** is oxidised by potassium manganate(VII), which excludes 2-hydroxy-2-methylpropanal, as it contains a tertiary alcohol that resists oxidation.

The oxidation product, **G**, has two degrees of unsaturation and a three-carbon skeleton. This suggests that **A** originally had both a secondary alcohol and an aldehyde, which upon oxidation, gave rise to a ketone and a carboxylic acid. Among the possible isomers, only 2-hydroxybutanal fits this description. Structures like 4-hydroxybutanal and 2-hydroxy-2-methylpropanal are ruled out as they contain primary alcohols.

Once 2-hydroxybutanal is identified as compound **A**, the structures of compounds **B** through **F** can be logically deduced. Special attention must be paid to how HCl adds across the C=C bond in **B** to ensure the final compound **F** does not include a –CH(OH)CH_3 group, which would otherwise yield a positive iodoform test. As expected, compound **E**, formed from the reduction of an aldehyde, is a primary alcohol and will not give a positive iodoform result.

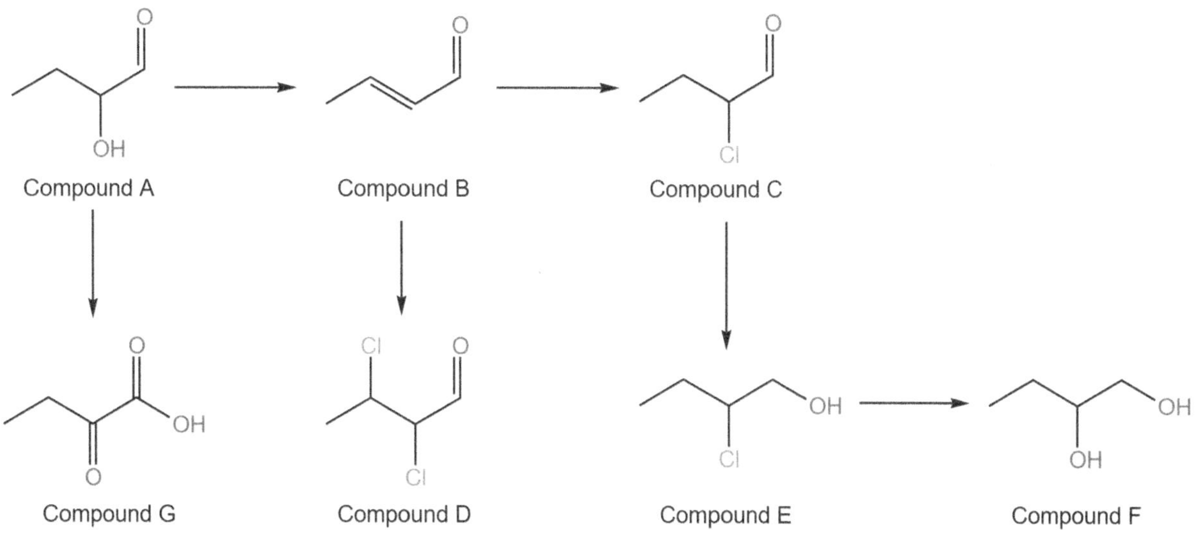

Question 39:

Compound **A** ($C_4H_8O_2$) produces an orange precipitate when treated with 2,4-dinitrophenylhydrazine and forms a silver mirror upon reaction with Tollen's reagent. When subjected to acidic conditions, **A** is transformed into compound **B** (C_4H_6O), which can exist as either a cis- or trans-isomer. Compound **B** reacts with hydrogen chloride to give compound **C** (C_4H_7OCl) and undergoes chlorination to produce compound **D** ($C_4H_6OCl_2$).

Reduction of compound **C** results in compound **E** (C_4H_9OCl), which is hydrolyzed by dilute alkali to form compound **F** ($C_4H_{10}O_2$). Compound **F** tests positive in the iodoform reaction (CHI_3), while compounds **A** and **E** do not. Strong oxidation of compound **A** gives compound **G** with the molecular formula $C_4H_6O_3$. Use this information to deduce the identities and structural formulas of compounds **A** through **G**, and explain your reasoning in detail.

My thought process:

Compound **A** has one degree of unsaturation, contains two oxygen atoms, and gives positive results with both 2,4-dinitrophenylhydrazine and Tollens' reagent. These observations point strongly toward **A** being a hydroxyaldehyde. Among the possible structures for a four-carbon compound containing both hydroxyl and aldehyde functional groups are: 2-hydroxybutanal, 3-hydroxybutanal, 4-hydroxybutanal, 2-hydroxy-2-methylpropanal, and 3-hydroxy-2-methylpropanal.

When **A** undergoes acid-catalysed dehydration, it forms compound **B**, which features both an alkene and an aldehyde group. The key clue here is that **B** shows geometrical isomerism, which suggests that A must contain a secondary or tertiary alcohol—since only these types can give rise to substituted alkenes capable of cis-trans isomerism. This rules out candidates with primary alcohols, such as 4-hydroxybutanal and 3-hydroxy-2-methylpropanal.

🔍 **Tip for students:** Only alkenes where both carbon atoms of the double bond are bonded to two different groups can show cis-trans isomerism. Hence, dehydration of a primary alcohol generally won't yield such a product.

The electrophilic additions of HCl and Cl_2 to **B** serve to confirm the presence of a C=C double bond. The aldehyde functionality remains intact in both **C** and **D**, which are subsequently reduced to give compound **E**—a primary alcohol that also contains a halogen. When **E** is treated with aqueous alkali, the halogen is substituted to form a diol, **F**.

To further narrow down the identity of **A**, note that it does not give a positive result in the iodoform test. This eliminates 3-hydroxybutanal, as it contains the –CH(OH)CH_3 group responsible for a positive result. Similarly, 2-hydroxy-2-methylpropanal can be excluded because it is a tertiary alcohol, which is resistant to oxidation, while **A** is clearly oxidisable by $KMnO_4$. The only viable structure left for **A** is 2-hydroxybutanal.

Attentive students may recognise that this question mirrors the previous one (Q38), with the exception that here, compound **F** does test positive for the iodoform test. Once **A** is confirmed as 2-hydroxybutanal, the identities of compounds B through F can be logically deduced. One must take care in determining how HCl adds across the C=C double bond in **B**, to ensure that the final diol, **F**, contains the $-CH(OH)CH_3$ group required to yield a positive iodoform result. As expected, compound **E**—formed by reducing the aldehyde—is a primary alcohol and will not show a positive iodoform test.

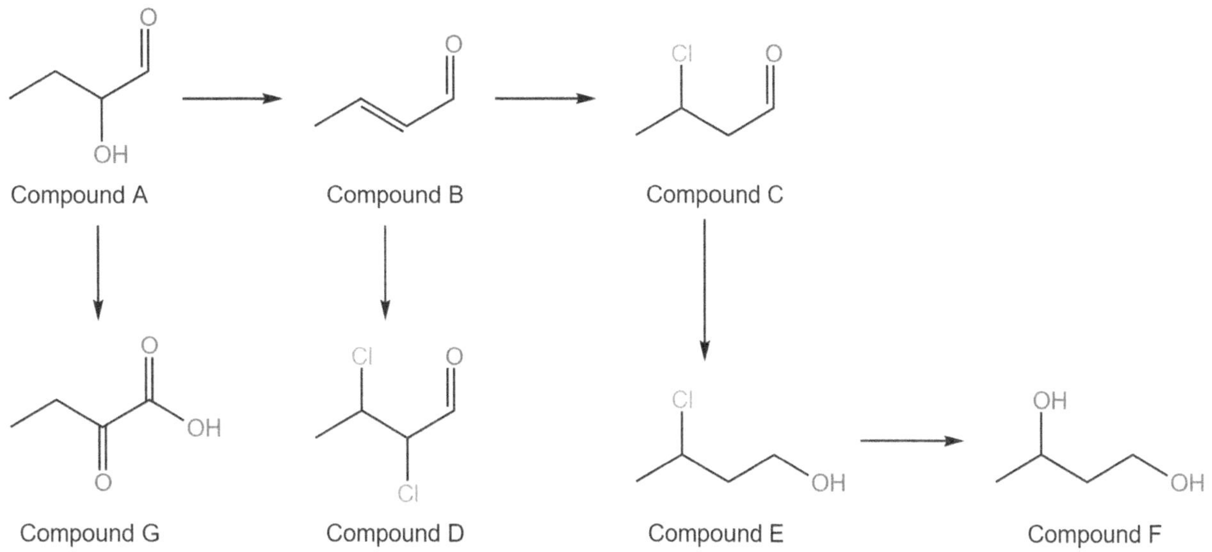

Question 40:

Compound **A** (C_3H_7Cl) reacts with aqueous sodium hydroxide, forming compound **C** (C_3H_8O). Treating **C** with phosphorus tribromide results in the release of steamy vapors. Oxidation of **C** using acidified potassium dichromate (VI) produces compound **D** (C_3H_6O), which is further oxidized to compound **E** ($C_3H_6O_2$). Compound **D** reacts positively with 2,4-dinitrophenylhydrazine and Tollen's reagent. Compound **E** reacts with sodium hydrogen carbonate, releasing carbon dioxide, and with phosphorus tribromide.

In the presence of benzene and anhydrous iron(III) chloride, **A** forms compound **B** (C_9H_{12}), whose structure is unexpected based on **A**. Interestingly, compound **B** can also be synthesized from benzene and **F**, an isomer of **A**, to produce the expected product. When treated with sodium hydroxide, **F** is converted into compound **G**, an isomer of **C**. Oxidizing **G** with acidified potassium dichromate (VI) results in compound **H**, an isomer of **D**, with no further oxidation possible. Compound **H** reacts positively with 2,4-dinitrophenylhydrazine, does not react with Tollen's reagent, and gives a positive result in the iodoform test.

Heating **B** with alkaline potassium permanganate (VII) yields an aqueous solution of **I**. Upon treatment with dilute sulfuric acid, **I** produces a white precipitate, identified as **J**.

(a) Identify compounds **A** through **J**, providing detailed explanations of the observations and supporting your reasoning with balanced chemical equations.

(b) Discuss why both **A** and **F** produce the same product when reacted with benzene.

My thought process:

Compound **A** has a degree of unsaturation of zero and contains only three carbon atoms. At this point, students are encouraged to quickly sketch all possible isomers using skeletal formulae to visualise their options. In this case, there are only two possible structures for **A**: 1-chloropropane or 2-chloropropane.

When **A** undergoes alkaline hydrolysis, it yields compound **C**, an alcohol. Reacting **C** with phosphorus tribromide (PBr_3) produces a corresponding alkyl bromide, confirming the presence of a hydroxyl group in compound **C**. Subsequent oxidation of **C** with acidified potassium dichromate ($K_2Cr_2O_7$) proceeds in two stages: first forming an aldehyde (compound **D**), then further oxidising to a carboxylic acid (compound **E**). This oxidation pattern is characteristic of a primary alcohol, which tells us that compound **C** must be a primary alcohol—meaning that compound **A** must be 1-chloropropane, the primary alkyl halide.

Compound **B** has a degree of unsaturation equal to four, indicating the presence of an aromatic ring. To generate **B**, compound **A** reacts with benzene in the presence of a Lewis acid catalyst (e.g. anhydrous $FeCl_3$) in what is described as a Friedel–Crafts alkylation. The expected product would be a straightforward alkylbenzene, but in this case, the observed product is not what would normally result from direct substitution.

This deviation can be explained by understanding the mechanism of electrophilic aromatic substitution. In the presence of $FeCl_3$, compound **A** generates a propyl carbocation ($CH_3CH_2CH_2^+$). However, this is a primary carbocation, which is relatively unstable. To form a more stable intermediate, a hydride shift occurs, producing a secondary carbocation ($CH_3CH^+CH_3$). This more stable species is the actual electrophile that reacts with benzene, leading to the unexpected substitution product.

Tip for students: When faced with a Friedel–Crafts alkylation question where the product does not match initial expectations, it is often a signal that carbocation rearrangement is at play. Recognising this possibility is essential to understanding and predicting the outcome of the reaction.

Once the identity of compound **A** has been established, and with an awareness of possible rearrangements in carbocation intermediates, the structures of compounds **B** through **J** can be determined with relative ease by applying standard reaction principles.

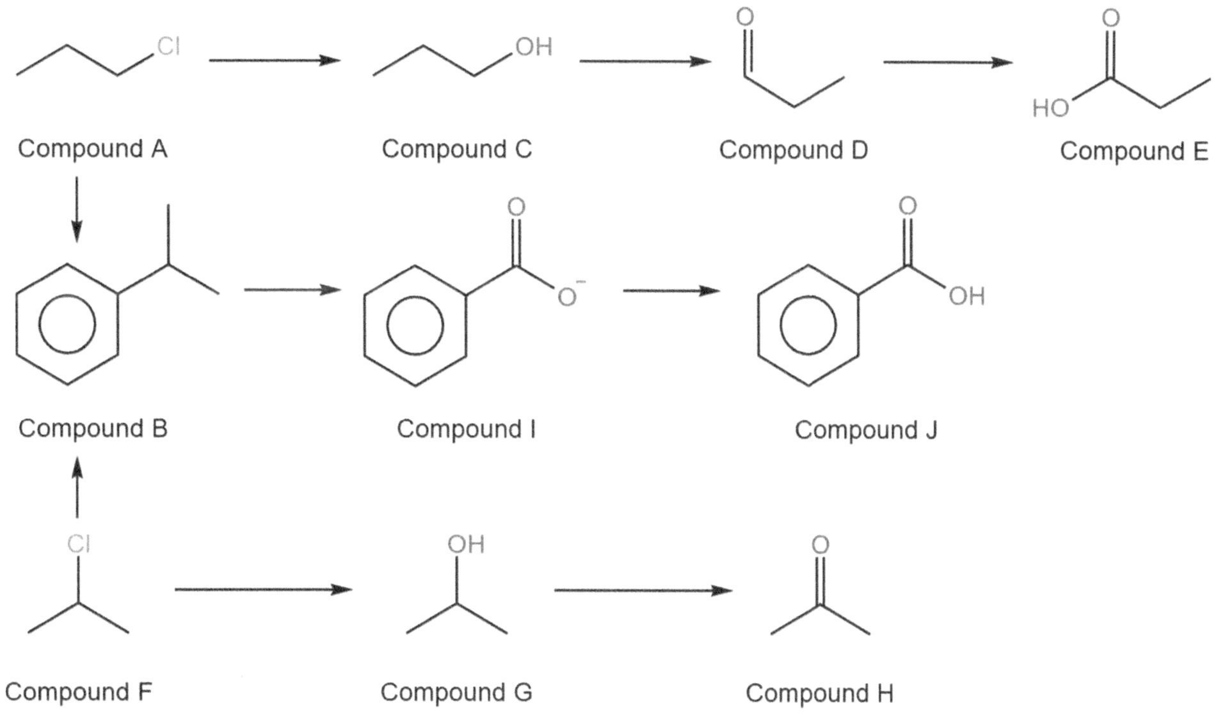

Question 41:

Compound **A** (C_4H_9Cl) undergoes a reaction with aqueous sodium hydroxide to form compound **C** ($C_4H_{10}O$). When treated with phosphorus pentachloride, **C** produces steamy fumes. Oxidation of **C** using acidified potassium dichromate (VI) results in the formation of compound **D** (C_4H_8O), which can be further oxidized to produce compound **E** ($C_4H_8O_2$). Compound **D** reacts positively with both 2,4-dinitrophenylhydrazine and Tollen's reagent. Compound **E**, when treated with sodium hydrogen carbonate solution, releases carbon dioxide and also reacts with phosphorus pentachloride.

In the presence of benzene and anhydrous aluminium chloride, compound **A** reacts to yield compound **B** ($C_{10}H_{14}$), which exhibits a structure differing from what would initially be anticipated based on **A**. Alternatively, compound **B** can be synthesized from benzene and **F**, an isomer of **A**, forming the expected product. Treating **F** with sodium hydroxide produces compound **G**, an isomer of **C**. Oxidizing **G** with acidified potassium dichromate (VI) results in compound **H**, an isomer of **D**, without further oxidation occurring. Compound **H** reacts positively with 2,4-dinitrophenylhydrazine, shows no reaction with Tollen's reagent, and produces a positive result in the iodoform test.

When compound **B** is heated with alkaline potassium permanganate (VII), it generates an aqueous solution of **I**. Upon treatment with dilute sulfuric acid, **I** forms a white precipitate identified as **J**.

(a) Using the provided information, determine the identities and structures of compounds **A** through **J**. Explain each observation comprehensively, supporting your reasoning with balanced chemical equations.

(b) Discuss the reason why both **A** and **F** yield the same product when reacted with benzene.

My thought process:

Compound **A** has a degree of unsaturation equal to zero, indicating that it is a saturated compound, most likely an alkyl chloride. When **A** undergoes alkaline hydrolysis, it yields alcohol **C**. Additionally, when **A** is treated with phosphorus pentachloride (PCl_5), the release of steamy fumes confirms the presence of a hydroxyl group, verifying that **C** is indeed an alcohol.

Further oxidation of **C** using acidified potassium dichromate ($K_2Cr_2O_7$) proceeds in two stages: first forming an aldehyde (**D**), followed by oxidation to a carboxylic acid (**E**). This two-step sequence is characteristic of primary alcohols, confirming that **C** is primary, and hence compound **A** must be a primary alkyl halide.

Given that **A** is a primary chloroalkane with four carbon atoms, students who are comfortable sketching skeletal structures should quickly recognise that the two possible isomers for **A** are 1-chlorobutane and 1-chloro-2-methylpropane.

Compound **B**, on the other hand, has a degree of unsaturation equal to four, which is a strong indicator of an aromatic ring, specifically a benzene ring. The transformation from **A** to **B** involves a Friedel–Crafts alkylation, where compound **A** acts as the alkylating agent. In the presence of anhydrous $AlCl_3$, **A** forms a primary carbocation ($CH_3CH_2CH_2CH_2^+$). However, because primary carbocations are inherently unstable, a hydride shift occurs to generate a more stable secondary carbocation ($CH_3CH_2CH^+CH_3$). It is this rearranged carbocation that reacts with benzene to form the final alkylbenzene product.

At this point, students might be unsure whether **A** is 1-chlorobutane or 1-chloro-2-methylpropane. The key to answering this lies in identifying the structure of compound B. When **B** is oxidised, it forms benzoic acid (**J**). This observation alone is sufficient to determine that **A** must be 1-chlorobutane, as its corresponding side chain is capable of being oxidised to a carboxylic acid when attached to a benzene ring.

To better understand the reasoning behind this conclusion—and to explore the alternative case—students are encouraged to attempt Question 42, which builds directly on this concept. The explanation for that case will clarify why only one of the two possible isomers fits the entire sequence of reactions.

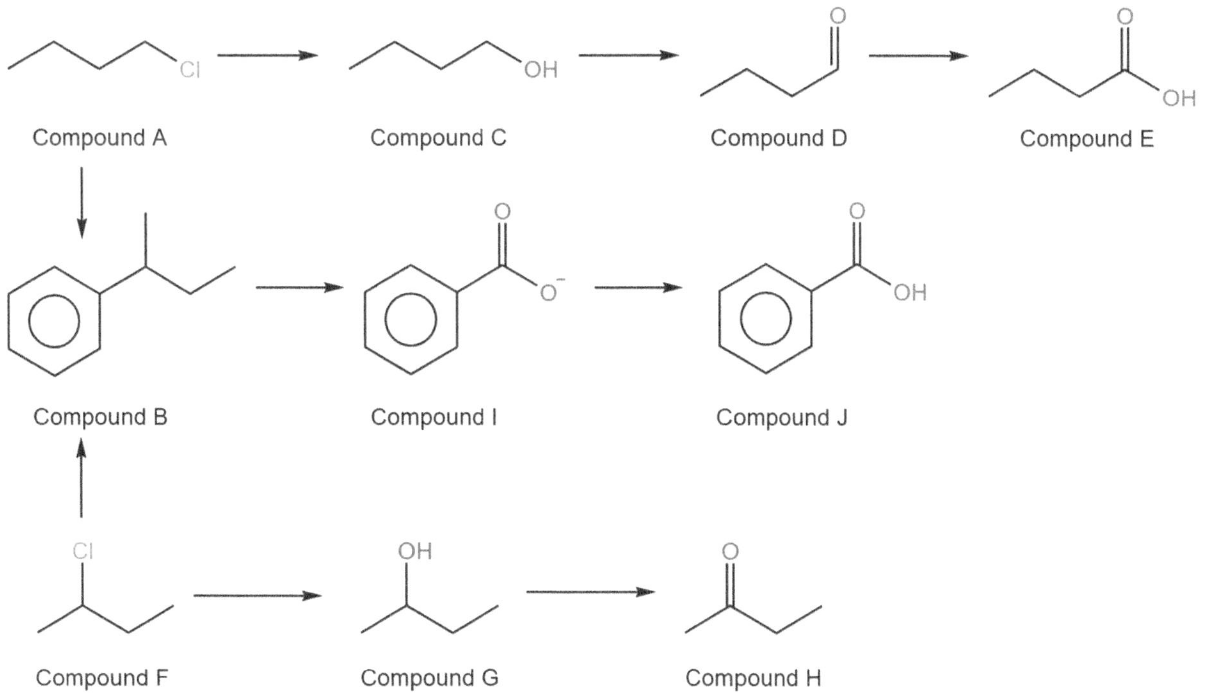

Question 42:

Compound **A** (C_4H_9Cl) undergoes a reaction with aqueous sodium hydroxide, producing compound **C** ($C_4H_{10}O$). Upon reaction with phosphorus pentachloride, **C** releases steamy fumes. Oxidizing **C** with acidified potassium dichromate (VI) yields compound **D** (C_4H_8O), which can be further oxidized to form compound **E** ($C_4H_8O_2$). Compound **D** exhibits positive responses with both 2,4-dinitrophenylhydrazine and Tollen's reagent, while compound **E** reacts with sodium hydrogen carbonate, releasing carbon dioxide, and also reacts with phosphorus pentachloride.

When reacted with benzene in the presence of anhydrous aluminium chloride, compound **A** forms **B** ($C_{10}H_{14}$), a structure different from what one might predict based on **A**. Compound **B** can alternatively be synthesized from benzene and compound **F**, an isomer of **A**, yielding the anticipated product. Treating **F** with sodium hydroxide produces compound **G**, an isomer of **C**. However, oxidation of **G** with acidified potassium dichromate (VI) is not possible. Additionally, compound **B** is resistant to oxidation by acidified potassium permanganate (VII).

(a) Using the details provided, identify the structures and identities of compounds **A** through **G**, explaining each observation thoroughly and supporting your reasoning with balanced equations.

(b) Offer an explanation as to why both **A** and **G** produce the same product when reacted with benzene.

My thought process:

This question closely mirrors the scenario in Q41, with one important difference: compounds **B** and **G** in this question cannot be oxidised. Like before, compound **A** is a primary alkyl halide with no degrees of unsaturation, which limits its possible structures to either 1-chlorobutane or 1-chloro-2-methylpropane.

Regardless of which isomer **A** is, both can undergo carbocation rearrangement in the presence of a Lewis acid catalyst, such as $AlCl_3$, to generate a more stable carbocation. In the case of 1-chlorobutane, the initially formed primary carbocation rearranges to a secondary carbocation ($CH_3CH_2CH^+CH_3$), as demonstrated in Q41. In contrast, 1-chloro-2-methylpropane rearranges directly to a highly stable tertiary carbocation ($C^+(CH_3)_3$).

In this particular question (Q42), the tertiary carbocation acts as the electrophile in a Friedel–Crafts alkylation, attaching to the benzene ring to form tert-butylbenzene as compound **B**. However, because this alkyl benzene lacks a benzylic hydrogen, it cannot undergo oxidation to a benzoic acid derivative. This stands in contrast to Q41, where the secondary alkylbenzene product does have a benzylic hydrogen, allowing it to be readily oxidised to benzoic acid.

This question is a great example of how a small structural difference—in this case, whether a benzylic hydrogen is present—can significantly influence the outcome of a reaction. It also reinforces why students are encouraged to practise drawing skeletal structures of isomers efficiently. Visualising possible isomers quickly is a valuable skill when working through multi-step organic questions like these, where structural detail is key to understanding the chemistry involved.

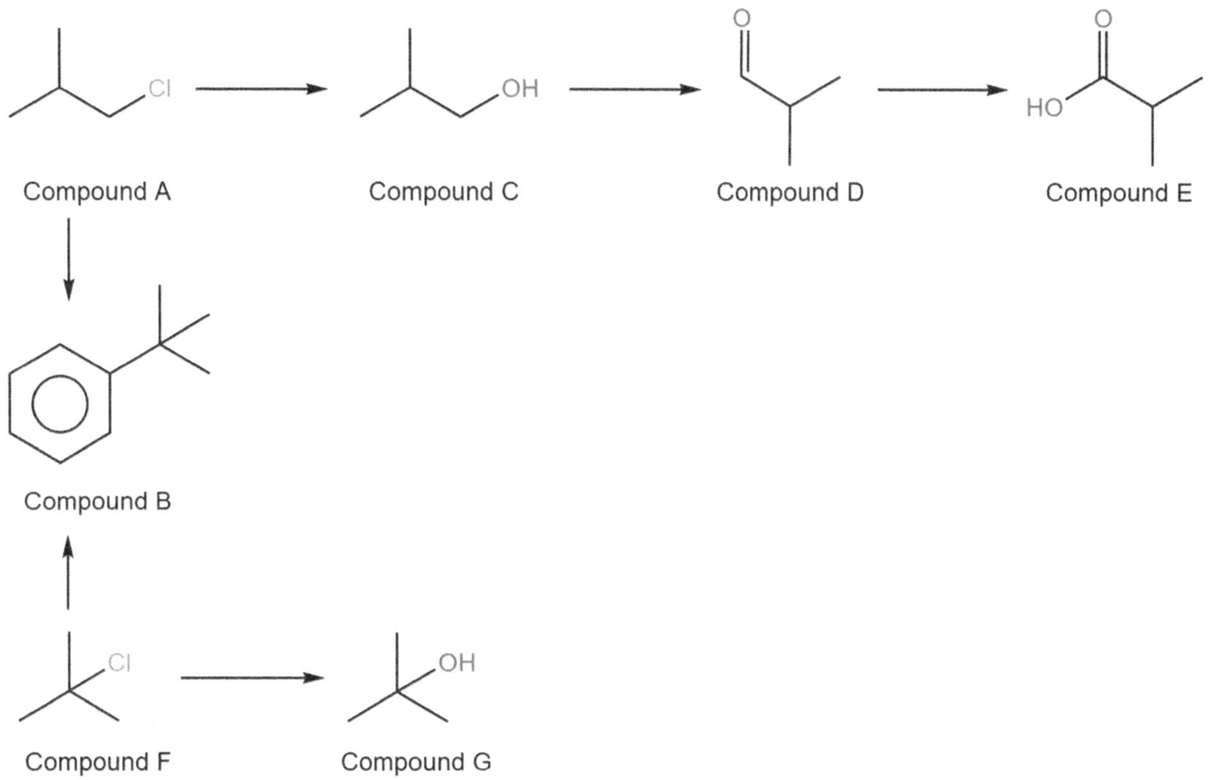

Question 43:

Compound **A** (C_4H_9Cl) reacts with aqueous sodium hydroxide to produce compound **C** ($C_4H_{10}O$). When treated with phosphorus pentachloride, **C** releases steamy fumes. However, compound **C** does not undergo oxidation when exposed to acidified potassium dichromate (VI).

Compound **A** reacts with benzene in the presence of anhydrous aluminium chloride to yield compound **B** ($C_{10}H_{14}$), which cannot be oxidized by acidified potassium permanganate (VII).

Identify the structures and compositions of compounds **A**, **B**, and **C**, and provide detailed explanations for the observations, including balanced equations to support your reasoning.

My thought process:

Compound **A** exhibits no degrees of unsaturation, indicating that it is a fully saturated molecule, most likely an alkyl chloride. Upon alkaline hydrolysis, **A** produces compound **C**, an alcohol. When **A** is treated with phosphorus pentachloride (PCl_5), the evolution of steamy fumes further confirms the presence of a hydroxyl group, supporting that **C** is indeed an alcohol.

However, compound **C** does not undergo oxidation with acidified potassium dichromate ($K_2Cr_2O_7$), which suggests that it is a tertiary alcohol. Given that **C** contains four carbon atoms, the only structure that fits this description is 2-methylpropan-2-ol. As a result, compound **A** must be 2-chloro-2-methylpropane, the corresponding tertiary alkyl chloride.

When **A** is reacted with benzene in the presence of anhydrous $AlCl_3$, it forms a tertiary carbocation. Since this carbocation is already at its most stable configuration, no rearrangement is needed. The electrophilic substitution proceeds smoothly, producing tert-butylbenzene as compound **B**.

Because tert-butylbenzene lacks a benzylic hydrogen, it is resistant to oxidation, and no further transformation into a carboxylic acid is possible under typical conditions.

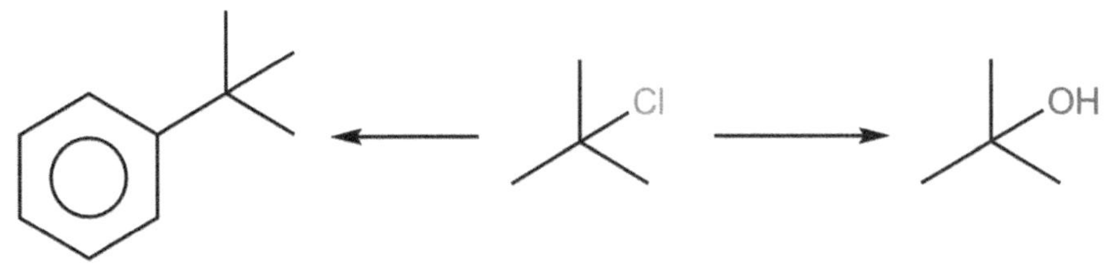

Compound B Compound A Compound C

Question 44:

Compound **A** ($C_5H_{11}Cl$) reacts with aqueous sodium hydroxide to form compound **B** ($C_5H_{12}O$). When **B** is treated with acidified potassium permanganate (VII), it is converted to compound **C**, which does not produce a yellow precipitate when warmed with alkaline aqueous iodine. Heating **B** with concentrated sulfuric acid at 170°C leads to the formation of compound **D** (C_5H_{10}). Vigorous oxidation of **D** using hot concentrated acidified potassium permanganate (VII) results in the formation of butanoic acid as the sole organic product.

Deduce the structural formulas of compounds **A**, **B**, **C**, and **D**, providing detailed reasoning for each step. Ensure all observations are explained thoroughly, supported by balanced equations.

My thought process:

Compound **A** has a degree of unsaturation equal to zero, indicating that it is a fully saturated alkyl chloride. Upon undergoing alkaline hydrolysis, **A** is converted into an alcohol, compound **B**. Since **B** is successfully oxidised by acidified potassium manganate(VII), it must be either a primary or secondary alcohol, as tertiary alcohols are resistant to oxidation under such conditions.

Tip for students: If an alcohol produces a positive iodoform test, it cannot be tertiary. Only one primary alcohol—ethanol—gives a positive result, and this is due to the presence of the –CH(OH)CH_3 moiety. For any alcohol with three or more carbon atoms, a positive result suggests the alcohol is secondary, not primary. A negative result does not, however, confirm the alcohol is primary; further tests are needed for confirmation.

When compound **B** is subjected to acid-catalysed dehydration, it forms an alkene, compound **D**. Oxidative cleavage of **D** using hot $KMnO_4$ produces only one organic product, identified as butanoic acid, a four-carbon carboxylic acid. This suggests that the original alkene D was terminal, with one carbon lost as CO_2 during cleavage—consistent with the structure of pent-1-ene.

Given that **D** is pent-1-ene, a linear terminal alkene, it follows that alcohol **B** must have been a straight-chain primary alcohol, specifically pentan-1-ol. Otherwise, if compound **B** had been pentan-2-ol, it would have produced a positive result in the iodoform (triiodomethane) test due to the presence of the –CH(OH)CH_3 group. Once **B** is confirmed, the identity of compound **A** (which must be 1-chloropentane) and compound **C** (pentanoic acid) can be readily deduced based on standard organic reactions.

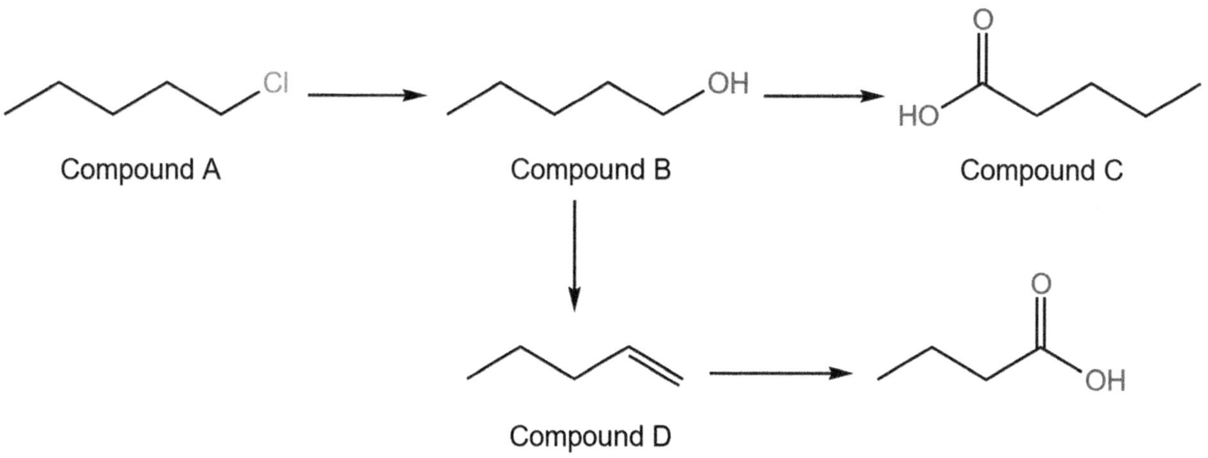
Cl
OH
O
HO
Compound A
Compound B
Compound C
Compound D
O
OH

Question 45:

Compound **A** ($C_5H_{11}Cl$) reacts with aqueous sodium hydroxide to produce compound **B** ($C_5H_{12}O$). Oxidation of **B** using acidified potassium permanganate (VII) forms compound **C**, which yields a yellow precipitate when warmed with alkaline aqueous iodine. Heating **B** with concentrated sulfuric acid at 170°C leads to the formation of two isomers, **D** and **E** (C_5H_{10}). Upon vigorous oxidation using hot, concentrated acidified potassium permanganate (VII), **D** produces both propanoic acid and ethanoic acid as organic products, while **E** exclusively forms butanoic acid.

Deduce the structural formulas of compounds **A**, **B**, **C**, **D**, and **E** based on the observations provided, and explain your reasoning comprehensively, including balanced chemical equations to support your answers.

My thought process:

Compound **A** exhibits no degrees of unsaturation, indicating it is a fully saturated alkyl chloride. When **A** undergoes alkaline hydrolysis, it forms an alcohol, compound **B**. The fact that **B** can be oxidised by acidified potassium manganate(VII) tells us it must be either a primary or secondary alcohol, as tertiary alcohols resist oxidation under these conditions.

Further evidence comes from **B**'s positive result in the triiodomethane (iodoform) test, which is characteristic of alcohols containing the $–CH(OH)CH_3$ group. Since **B** has more than two carbon atoms, it cannot be ethanol (the only primary alcohol that gives this test), and therefore B must be a secondary alcohol with the appropriate structural feature.

Dehydration of **B** results in the formation of two distinct alkenes, **D** and **E**. This is typical for secondary alcohols, as elimination of water can occur in either direction, producing more than one positional isomer of the alkene. Oxidative cleavage of compound **D**, a five-carbon alkene, yields propanoic acid and ethanoic acid. Because there is no net loss or gain in carbon atoms, **D** must be an internal alkene, identified as pent-2-ene. Whether cis or trans, the distinction is not relevant to this analysis.

On the other hand, oxidative cleavage of compound **E** produces only butanoic acid, a four-carbon carboxylic acid, confirming that **E** is a terminal alkene—specifically, pent-1-ene. Since **B** leads to the formation of both pent-1-ene and pent-2-ene through dehydration and is known to be a secondary alcohol, its structure must be pentan-2-ol.

Accordingly, compound **A**, the precursor alkyl chloride, must be 2-chloropentane, and compound **C**—formed by oxidation of **B**—is pentan-2-one, a ketone consistent with the oxidation of a secondary alcohol.

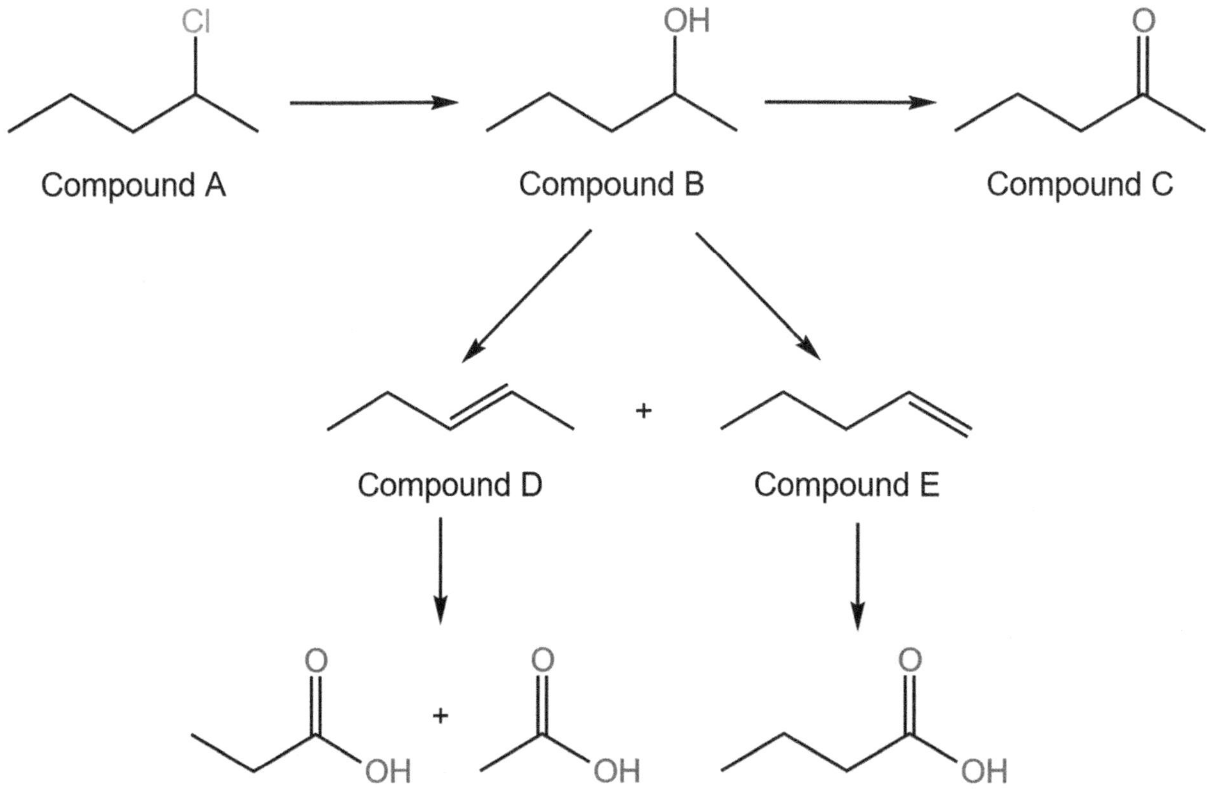
Cl
OH
O
Compound A
Compound B
Compound C
+
Compound D
Compound E
O
OH
+
O
OH
O
OH

Question 46:

Compound **A** ($C_5H_{11}Cl$) reacts with aqueous sodium hydroxide, forming compound **B** ($C_5H_{12}O$). Oxidation of **B** using acidified potassium permanganate (VII) results in the formation of compound **C**, which does not produce a yellow precipitate when warmed with alkaline aqueous iodine. When **B** is heated with concentrated sulfuric acid at 170°C, compound **D** (C_5H_{10}) is formed, which may exist as cis- and trans-isomers. Vigorous oxidation of **D** with hot concentrated acidified potassium permanganate (VII) produces propanoic acid and ethanoic acid as the only organic products.

Deduce the structural formulas for compounds **A**, **B**, **C**, and **D**, providing detailed explanations for your conclusions and supporting your reasoning with balanced chemical equations.

My thought process:

Compound **A** exhibits no degrees of unsaturation, indicating that it is a saturated alkyl chloride. When **A** undergoes alkaline hydrolysis, it yields compound **B**, an alcohol. Since **B** is readily oxidised by acidified potassium manganate(VII), it must be either a primary or secondary alcohol, as tertiary alcohols resist oxidation under these conditions.

At this stage, without molecular data for compound **C** (the oxidation product of B), it is unclear whether **C** is a ketone or a carboxylic acid. However, one thing is certain—compound **C** does not contain a methyl ketone group ($-COCH_3$), as no iodoform test result is noted.

Dehydration of alcohol **B** yields two geometrical isomers, suggesting that the resulting alkene **D** is an internal alkene, capable of exhibiting cis–trans isomerism. Further evidence comes from the oxidative cleavage of **D** using hot $KMnO_4$, which produces propanoic acid and ethanoic acid, while retaining the original number of carbon atoms. From this, it can be deduced that alkene **D** is pent-2-ene.

Once **D** is established, attention turns to the hydration of the C=C double bond. Water must have added across the double bond in such a way that the resulting alcohol **B** does not contain the $-CH(OH)CH_3$ group, which would otherwise give a positive iodoform test. Since **B** does not give a positive result, it must be pentan-3-ol, a secondary alcohol that lacks the necessary moiety for iodoform reactivity.

🔍 **Tip for Students:** If an alcohol gives a positive iodoform (tri-iodomethane) test, it contains the $-CH(OH)CH_3$ group. Upon oxidation, this group is converted into a methyl ketone ($-COCH_3$), which also gives a positive iodoform test. Therefore, if the oxidised product of an alcohol (such as a ketone or aldehyde) does not give a positive iodoform test, then the original alcohol must not have contained the $-CH(OH)CH_3$ group. Always check both the alcohol and its oxidation product for consistency—if one gives a positive test, the other should too, unless the reactive group has been structurally altered or lost during the transformation.

Accordingly, compound **A** is identified as 3-chloropentane, compound **B** as pentan-3-ol, and compound **C** as its oxidised product, pentan-3-one.

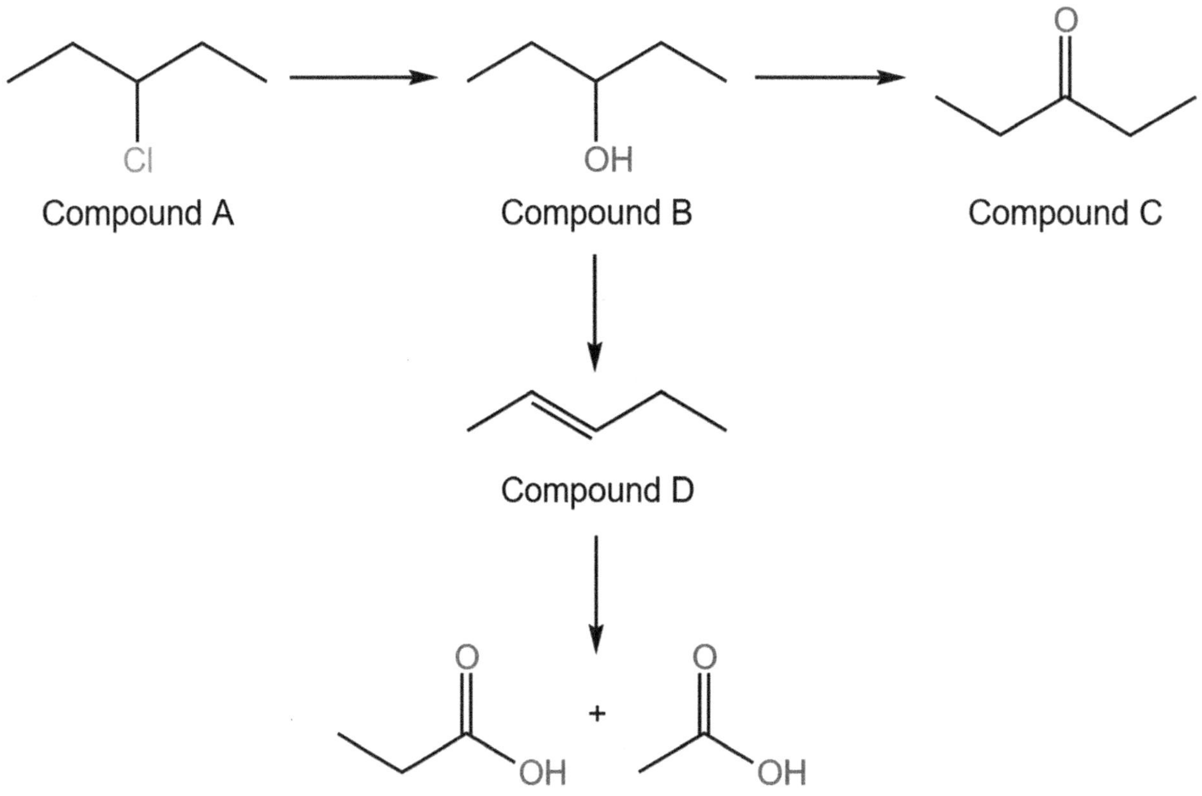

Question 47:

Compound **A** ($C_5H_{11}Cl$) reacts with aqueous sodium hydroxide to produce compound **B** ($C_5H_{12}O$). When subjected to oxidation with acidified potassium permanganate (VII), **B** forms compound **C**, which does not produce a yellow precipitate upon warming with alkaline aqueous iodine. Heating **B** with concentrated sulfuric acid at 170°C results in the formation of compound **D** (C_5H_{10}). Vigorous oxidation of **D** with hot concentrated acidified potassium permanganate (VII) exclusively yields butanone as the organic product.

Deduce the structural formulas for compounds **A**, **B**, **C**, and **D** based on the provided information. Provide detailed reasoning for each step and ensure observations are explained thoroughly, supported by balanced chemical equations.

My thought process:

Compound **A** has no degrees of unsaturation, indicating it is a fully saturated alkyl chloride. Upon alkaline hydrolysis, **A** is converted into compound **B**, an alcohol. Since **B** can be oxidised by acidified potassium manganate(VII), it must be either a primary or secondary alcohol, as tertiary alcohols are resistant to oxidation under these conditions. At this point, the precise identity of compound **C** (the oxidation product of B) remains uncertain—whether it is a ketone or a carboxylic acid—as no molecular formula is provided. However, what can be confirmed is that **C** does not contain a methyl ketone group ($-COCH_3$), as there is no evidence of a positive iodoform test.

When alcohol **B** is subjected to acid-catalysed dehydration, it forms compound **D**, an alkene. The oxidative cleavage of **D** using hot potassium manganate(VII) produces only one organic product: butanone, a four-carbon ketone. This suggests that one carbon atom has been lost during oxidation, which is characteristic of a terminal alkene undergoing cleavage. From this, compound **D** can be identified as 2-methylbut-1-ene.

There are two possible ways that water can be added across the C=C double bond of 2-methylbut-1-ene to yield an alcohol: either via Markovnikov or anti-Markovnikov addition. Only the anti-Markovnikov pathway leads to the formation of a primary alcohol—specifically, 2-methylbutan-1-ol, which matches the observed oxidation behaviour of **B**. The Markovnikov addition, by contrast, would give a tertiary alcohol (2-methylbutan-2-ol), which would not be oxidised under the same conditions.

It's important to note that in structure elucidation problems, one should not assume Markovnikov addition by default. Always confirm your reasoning with the clues given in the question, as relying solely on common reaction trends may lead to incorrect conclusions. In this context, the term "Markovnikov or anti-Markovnikov addition" is used simply to explain the logic behind selecting the correct structure for **B**, and not to imply that the reaction must proceed through this mechanism.

Once compound **B** is identified as 2-methylbutan-1-ol, it naturally follows that compound **A** is 1-chloro-2-methylbutane, and compound **C**, its oxidation product, is 2-methylbutanoic acid.

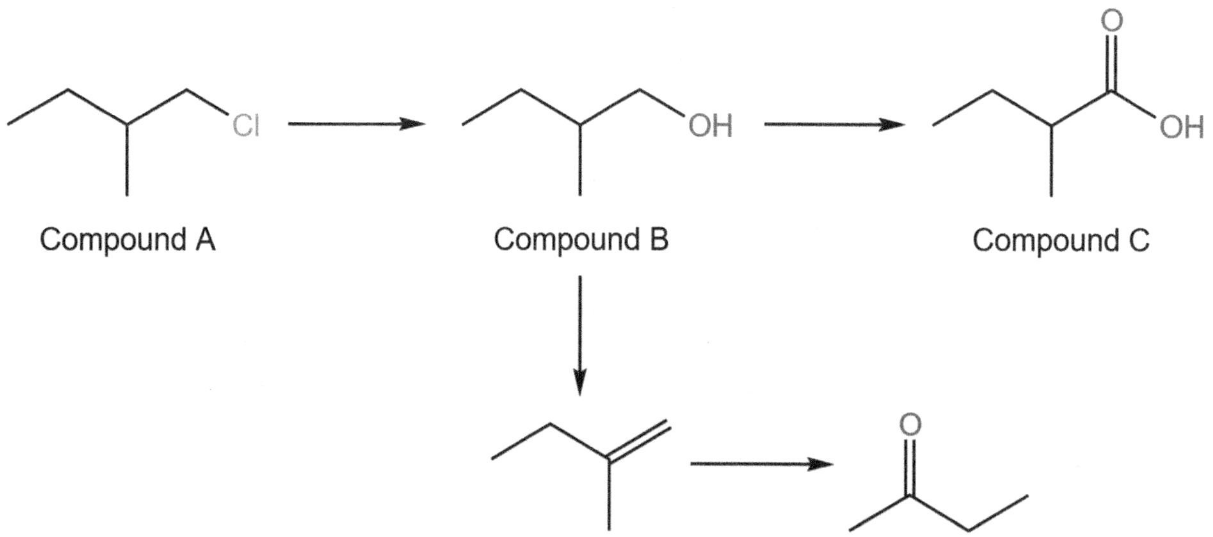

Compound A Compound B Compound C

Compound D

Question 48:

Compound **A** ($C_5H_{11}Cl$) reacts with aqueous sodium hydroxide to yield compound **B** ($C_5H_{12}O$). No observable reaction occurs when compound **B** is treated with acidified potassium permanganate (VII). Heating **B** with concentrated sulfuric acid at 170°C results in the formation of two isomers, **C** and **D** (C_5H_{10}). When subjected to vigorous oxidation using hot concentrated acidified potassium permanganate (VII), **C** produces butanone as the sole organic product, whereas **D** yields ethanoic acid and propanone.

Determine the structural formulas for compounds **A**, **B**, **C**, and **D**, providing a detailed explanation of your reasoning. Ensure observations are clearly justified and supported by balanced chemical equations.

My thought process:

Compound **A** has no degrees of unsaturation, indicating that it is a saturated alkyl chloride. Upon alkaline hydrolysis, it produces compound **B**, an alcohol. The fact that **B** does not undergo oxidation with acidified potassium manganate(VII) suggests that it is a tertiary alcohol, as these are resistant to oxidation under standard conditions. Given that **B** contains five carbon atoms and is a tertiary alcohol, its structure can only be 2-methylbutan-2-ol. Once **B** is identified, the structures of compound **A**, as well as subsequent compounds **C** and **D**, can be determined with ease based on standard reaction pathways.

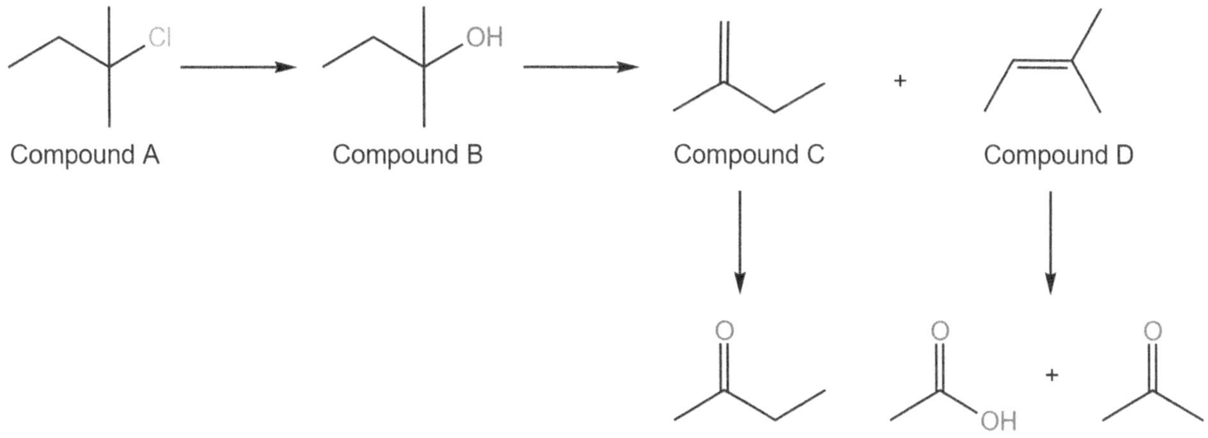

Question 49:

Compound **A** ($C_5H_{11}Cl$) reacts with aqueous sodium hydroxide to form compound **B** ($C_5H_{12}O$). Oxidation of **B** with acidified potassium permanganate (VII) yields compound **C**, which forms a yellow precipitate when warmed with alkaline aqueous iodine. Heating **B** at 170°C with concentrated sulfuric acid produces two isomers, **D** and **E** (C_5H_{10}). When subjected to vigorous oxidation with hot concentrated acidified potassium permanganate (VII), **D** produces 2-methylpropanoic acid as the sole organic product, while **E** results in the formation of ethanoic acid and propanone.

Determine the structural formulas for compounds **A, B, C, D,** and **E** using the given observations. Provide detailed explanations for each step and back up your reasoning with balanced chemical equations.

My thought process:

Compound **A** has no degrees of unsaturation, suggesting it is a fully saturated alkyl chloride. Upon alkaline hydrolysis, it forms an alcohol, compound **B**. Since **B** is readily oxidised by acidified potassium manganate(VII) and gives a positive iodoform (triiodomethane) test, it must be a secondary alcohol containing the characteristic –$CH(OH)CH_3$ group. This further implies that its oxidation product, compound **C**, is a methyl ketone.

When B undergoes acid-catalysed dehydration, it produces two different alkenes, **D** and **E**. The oxidative cleavage of alkene **D** (a five-carbon compound) yields 2-methylpropanoic acid, a four-carbon carboxylic acid. This outcome indicates that **D** is a terminal alkene, which can be identified as 3-methylbut-1-ene. In contrast, alkene **E** undergoes oxidative cleavage to form ethanoic acid and propanone, while maintaining the total carbon count. This result is consistent with the structure of 3-methylbut-2-ene.

The only alcohol that could give rise to both of these alkenes upon dehydration is 3-methylbutan-2-ol, a secondary alcohol that contains the –$CH(OH)CH_3$ moiety, aligning perfectly with the observed positive iodoform test. Once compound **B** is identified as 3-methylbutan-2-ol, the structures of compound **A** (its precursor alkyl chloride) and compound **C** (its oxidation product) can be deduced with confidence based on standard transformations.

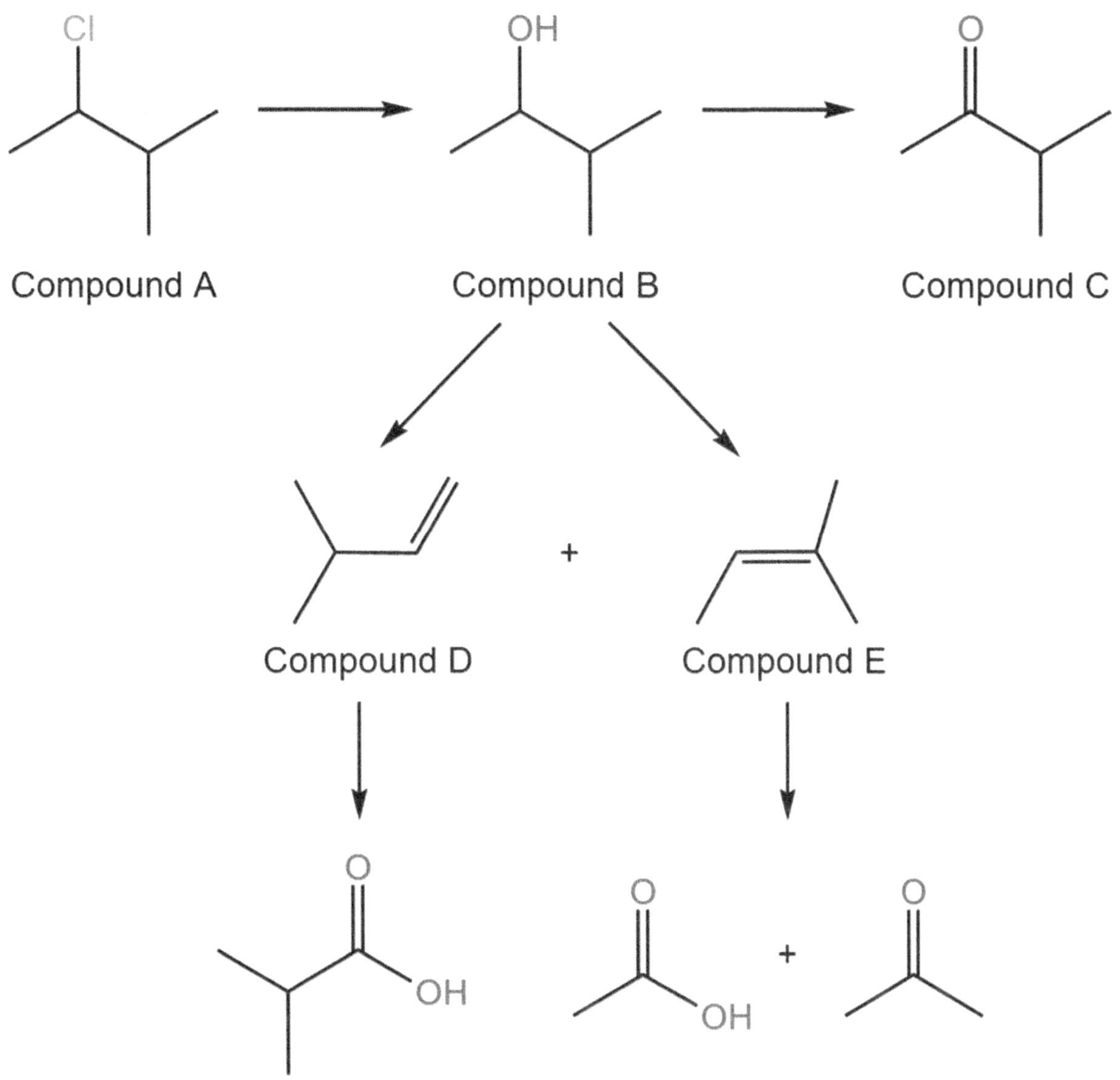
Cl
OH
O
Compound A
Compound B
Compound C
+
Compound D
Compound E
O
OH
O
OH
+
O

Question 50:

Compound **A** ($C_5H_{11}Cl$) reacts with aqueous sodium hydroxide, resulting in the formation of compound **B** ($C_5H_{12}O$). Oxidation of **B** using acidified potassium permanganate (VII) produces compound **C**, which does not generate a yellow precipitate when warmed with alkaline aqueous iodine. When **B** is heated with concentrated sulfuric acid at 170°C, it gives rise to compound **D** (C_5H_{10}). Vigorous oxidation of **D** with hot concentrated acidified potassium permanganate (VII) leads to the exclusive formation of 2-methylpropanoic acid as the organic product.

Deduce the structural formulas of compounds **A**, **B**, **C**, and **D** based on the information provided. Ensure that all reasoning is detailed and supported by balanced chemical equations.

My thought process:

Compound **A** shows no degree of unsaturation, indicating that it is a saturated alkyl chloride. When treated with aqueous alkali, **A** undergoes hydrolysis to produce alcohol **B**. The fact that B is oxidised by acidified potassium manganate(VII) suggests it is either a primary or secondary alcohol, as tertiary alcohols are resistant to oxidation.

However, compound **C**—the oxidation product of **B** (either a ketone or a carboxylic acid)—does not give a positive iodoform (triiodomethane) test, which tells us that alcohol **B** does not contain the –$CH(OH)CH_3$ moiety.

Interestingly, when alcohol **B** undergoes acid-catalysed dehydration, it forms only one alkene, compound **D**.

Tip for students: If dehydration of an alcohol yields only one alkene, it often means that either the alcohol is structurally symmetrical, or it is a primary alcohol with only one available elimination pathway. This is a useful clue in structure elucidation questions. Always pair it with other data (like oxidation behaviour or iodoform results) to confirm your deduction.

The oxidative cleavage of alkene **D** (which contains five carbon atoms) results in only one organic product, identified as 2-methylpropanoic acid (a four-carbon acid). This strongly suggests that alkene **D** is a terminal alkene, specifically 3-methylbut-1-ene.

From this, it follows that alcohol **B** must be 3-methylbutan-1-ol, which is a primary alcohol capable of undergoing oxidation with $KMnO_4$ but does not give a positive iodoform test. Once **B** is established, the identity of compound **A** (its precursor alkyl chloride) and compound **C** (its oxidation product - a ketone) can be determined accordingly.

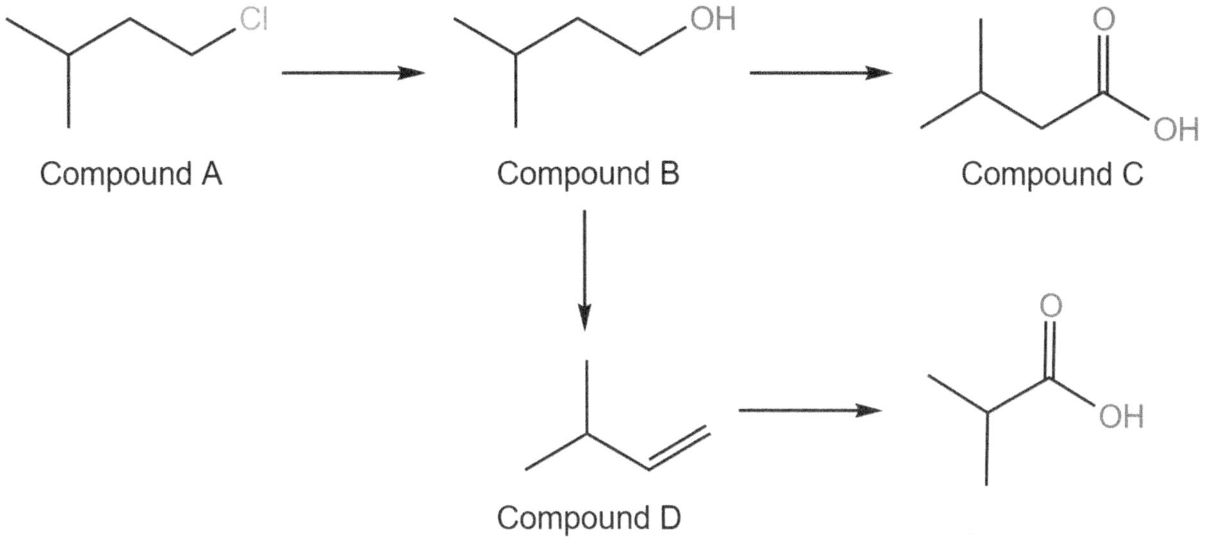
Cl
OH
O
OH
Compound A
Compound B
Compound C
O
OH
Compound D

Question 51:

Compound **A** ($C_5H_{11}Cl$) undergoes a reaction with aqueous sodium hydroxide, resulting in the formation of compound **B** ($C_5H_{12}O$). Upon oxidation with acidified potassium permanganate (VII), compound **B** is converted to compound **C**. Compound **C** does not yield a yellow precipitate when warmed with alkaline aqueous iodine. When **B** is heated at 170°C in the presence of concentrated sulfuric acid, no compound is produced.

Deduce the structural formulas of compounds **A**, **B**, and **C**, providing a detailed explanation of your reasoning and supporting your conclusions with relevant balanced chemical equations.

My thought process:

Compound **A** has no degree of unsaturation, suggesting that it is a saturated alkyl chloride. When **A** undergoes alkaline hydrolysis, it produces alcohol **B**. The fact that B is readily oxidised by acidified potassium manganate(VII) indicates it is either a primary or secondary alcohol. However, since compound **C** (the oxidation product of **B**) does not give a positive iodoform test, alcohol **B** does not possess the $-CH(OH)CH_3$ group.

An attempt to dehydrate alcohol **B** results in no alkene formation, which suggests that there is no hydrogen atom on the carbon adjacent to the hydroxyl-bearing carbon—a necessary condition for elimination to occur. Given that **B** is a five-carbon alcohol, the only structure that fits this description is 2,2-dimethylpropan-1-ol, where the central carbon is fully substituted and lacks adjacent β-hydrogens.

Once the structure of alcohol **B** is confirmed, the corresponding structures of compound **A** (its alkyl chloride precursor) and compound **C** (its oxidised product - a carboxylic acid) can be logically deduced.

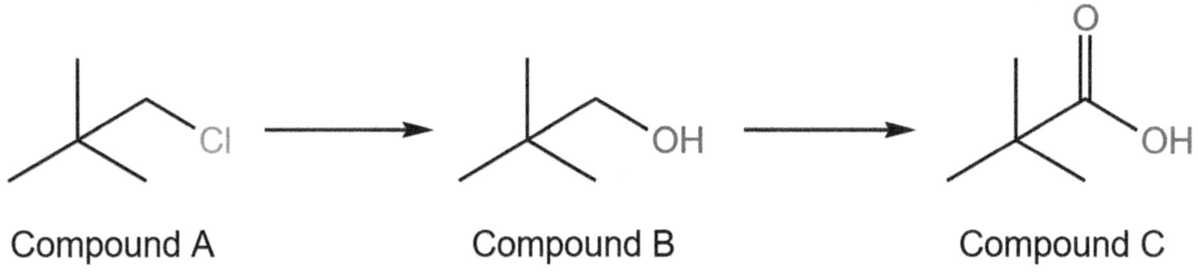

Question 52:

Compound **A** (C_2H_4) reacts with hydrogen bromide in the gaseous phase to produce compound **B** (C_2H_5Br). When **B** reacts with aqueous sodium hydroxide, it forms compound **C** (C_2H_6O). Oxidation of **C** with acidified potassium dichromate successively yields compound **D** (C_2H_4O) and then **E** ($C_2H_4O_2$). Compound **D** reacts positively with both 2,4-dinitrophenylhydrazine and Tollen's reagent.

By reacting **B** with magnesium in dry ether, compound **F** (C_2H_5MgBr) is formed. Compound **F** reacts with **D**, yielding compound **G** ($C_4H_{10}O$). Oxidation of **G** with acidified potassium dichromate produces compound **H** (C_4H_8O), which can further react with **F** to form compound **I** ($C_6H_{14}O$). Compound **I** does not undergo oxidation when treated with acidified potassium dichromate.

When compound **A** reacts with bromine in the presence of aqueous sodium chloride, two products are obtained: compound **J** ($C_2H_4Br_2$) and compound **K** (C_2H_4BrCl). Notably, no $C_2H_4Cl_2$ is formed during this reaction.

When **K** is treated with a limited quantity of aqueous sodium hydroxide, compound **L** (C_2H_5OCl) is produced as the major product.

(a) Identify the structural formulas and compositions of compounds **A** through **L** based on the information provided, and explain the observations comprehensively with supporting balanced equations.

(b) Explain why no $C_2H_4Cl_2$ is formed in the reaction described in the second-to-last paragraph.

(c) Justify why C_2H_5OCl is favored over C_2H_5OBr as the major product in the earlier mentioned reaction.

My thought process:

(a) Unlike previous questions where the identity of the starting compound **A** had to be deduced through a series of clues and backward reasoning, this question presents a much more straightforward scenario. It is immediately apparent that compound **A** is ethene, a simple alkene.

Ethene (**A**) undergoes electrophilic addition with hydrogen bromide, producing bromoethane (**B**). Compound **B** then undergoes alkaline hydrolysis, giving ethanol (**C**). Being a primary alcohol, ethanol can be oxidised to ethanal (**D**), and with further oxidation, to ethanoic acid (**E**). Since **D** is an aldehyde, it is expected to give positive tests with both 2,4-DNPH and Tollen's reagent.

Bromoethane (**B**) can also be reacted with magnesium in dry ether, forming a Grignard reagent, C_2H_5MgBr (**F**). This step is known as oxidative insertion, which effectively converts the electrophilic bromoethane into a nucleophilic species. In essence, this reagent behaves like a carbanion ($C_2H_5^-$), making it extremely useful for nucleophilic addition and nucleophilic substitution reactions.

When Grignard reagent **F** is reacted with ethanal (**D**), it undergoes nucleophilic addition to form compound **G**, a secondary alcohol. Understanding this mechanism is essential for students, as it enables them to accurately deduce the product structure by tracking electron flow.

Tip for students: A major advantage of using Grignard reagents is their ability to build carbon chains in a "step-up" fashion, adding multiple carbon atoms in a single step. This contrasts with methods like cyanide addition, which adds one carbon at a time. In synthetic design, minimising the number of steps is critical, as each step often leads to product loss, reducing overall yield.

Once the structure of alcohol **G** is established, identifying compound **H** becomes straightforward—it is the oxidation product of G. Reacting Grignard reagent **F** with **H** in a second round of nucleophilic addition gives compound **I**, which contains a tertiary alcohol structure.

(b) This question explores a variation on a familiar electrophilic addition mechanism, involving the reaction of ethene with bromine (Br_2) in the presence of aqueous sodium chloride (NaCl). In aqueous conditions, NaCl dissociates completely, providing free chloride ions (Cl^-) in solution.

The key step begins when the π-electrons of ethene (the alkene) act as a nucleophile and approach Br_2, which behaves as an electrophile. This electron-rich double bond polarises the Br–Br bond, weakening it and leading to heterolytic cleavage. The result is the formation of a bromonium ion intermediate, a three-membered cyclic structure with a positively charged bromine atom.

It is important to emphasise that water does not act as the first electrophile in this process. This is because the electron cloud on the oxygen atom in water is tightly held, making it difficult to polarise compared to the more diffuse electron cloud of Br_2, which is more easily polarised. Therefore, it is Br_2—not water—that first reacts with the alkene.

Several products can form from this mechanism:

- **J**: 1,2-dibromoethane ($C_2H_4Br_2$), when a bromide ion (Br^-) opens the bromonium ion.

- **K**: 1-bromo-2-chloroethane (C_2H_4BrCl), formed when a chloride ion (Cl^-) acts as the nucleophile.

- It is also theoretically possible to obtain 2-bromoethanol (C_2H_4BrOH) if water acts as the nucleophile, though this product is not the focus of the question.

Importantly, 1,2-dichloroethane ($C_2H_4Cl_2$) is not formed because chlorine gas (Cl_2) is not present. The only available chlorine species is Cl^-, which is a nucleophile, not an electrophile. Since electrophilic addition must be initiated by an electrophile, Cl^- cannot begin the reaction. Therefore, bromine is always the first species to add across the double bond.

(c) When compound **K** (1-bromo-2-chloroethane) is treated with aqueous sodium hydroxide, both the bromine and chlorine atoms are theoretically subject to nucleophilic substitution. However, because only a limited amount of NaOH is used, substitution occurs preferentially at the bromine site, as the C–Br bond is weaker and more reactive than the C–Cl bond. As a result, the monohydroxy compound becomes the major product.

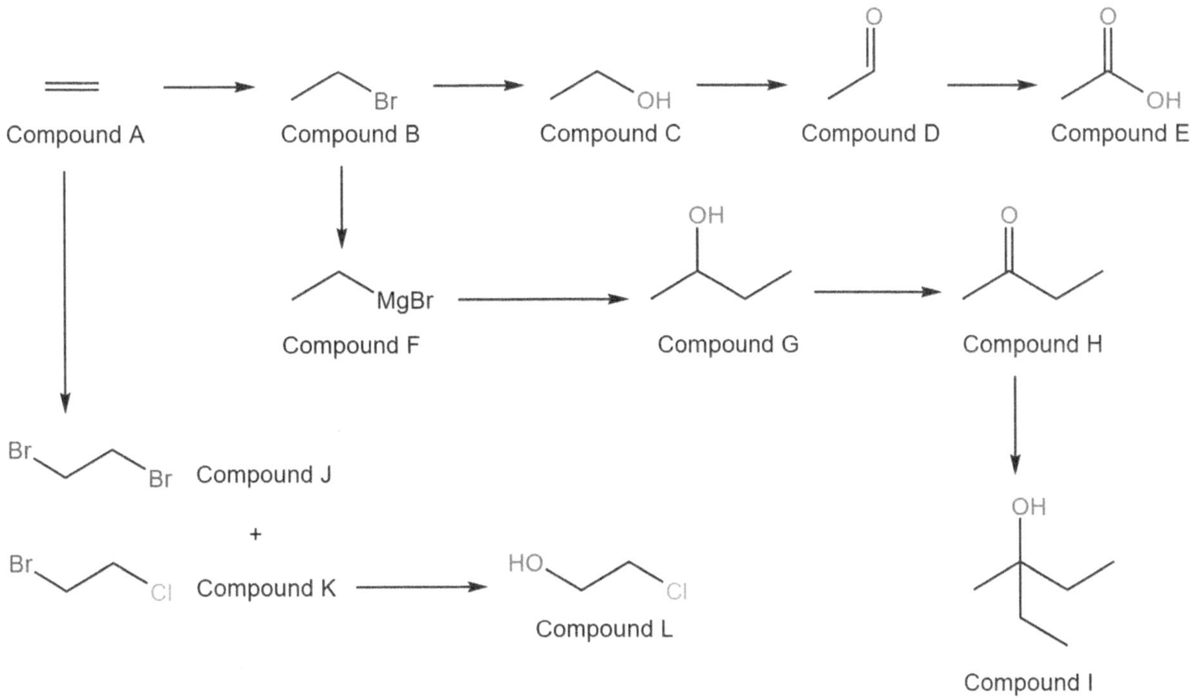

Question 53:

Compound **A** (C_2H_4) undergoes a reaction with hydrogen bromide in the gaseous phase, forming compound **B** (C_2H_5Br). Compound **B** reacts with aqueous sodium hydroxide to yield compound **C** (C_2H_6O), which can then undergo oxidation with acidified potassium dichromate to successively form compound **D** (C_2H_4O) and compound **E** ($C_2H_4O_2$). Compound **D** shows positive reactions with both 2,4-dinitrophenylhydrazine and Tollen's reagent.

Reaction of compound **B** with magnesium in dry ether produces compound **F** (C_2H_5MgBr), which reacts with **D** to give compound **G** ($C_4H_{10}O$). Oxidizing **G** with acidified potassium dichromate produces compound **H** (C_4H_8O), which can further react with **F**, yielding compound **I** ($C_6H_{14}O$). Compound **I** cannot be oxidized when treated with acidified potassium dichromate.

When compound **A** is treated with aqueous bromine, two products are formed: compound **J** (C_2H_4OBr) and compound **K** ($C_2H_4Br_2$). It is notable that compound $C_2H_6O_2$ is not produced in this reaction.

(a) Deduce the identities and structures of compounds **A** through **K**, and provide detailed explanations for each observation, supported by balanced equations where applicable.

(b) Explain the absence of $C_2H_6O_2$ in the reaction involving compound **A** and aqueous bromine.

My thought process:

It is immediately clear that compound **A** is ethene. Ethene undergoes electrophilic addition with hydrogen bromide (HBr) to form compound **B**, bromoethane. When **B** is treated with aqueous alkali, it undergoes nucleophilic substitution to give compound **C**, ethanol. Ethanol is a primary alcohol, and under oxidation with acidified potassium dichromate, it is converted first to ethanal (compound **D**) and, with further oxidation, to ethanoic acid (compound E). Since **D** is an aldehyde, it gives positive results with both 2,4-dinitrophenylhydrazine (2,4-DNPH) and Tollen's reagent.

In a separate reaction pathway, bromoethane (**B**) reacts with magnesium in dry ether, forming compound **F**, a Grignard reagent (C_2H_5MgBr). This Grignard reagent behaves as an ethyl nucleophile, which undergoes nucleophilic addition to the carbonyl group in ethanal (**D**), resulting in the formation of alcohol **G**. Students familiar with the mechanism of nucleophilic addition to aldehydes should be able to determine the structure of **G** with ease. Upon oxidation, alcohol **G** is converted into a ketone **H**, which can react once more with the Grignard reagent (**F**) to form compound **I**, a tertiary alcohol that is resistant to further oxidation by potassium dichromate ($K_2Cr_2O_7$).

Ethene (**A**) also reacts with aqueous bromine via electrophilic addition. In this mechanism, the π-electrons of ethene polarise the Br–Br bond, causing heterolytic cleavage and the formation of a bromonium ion intermediate. This highly reactive intermediate is then attacked by a nucleophile—in this example water or Br^-.

It is worth clarifying why a diol such as $C_2H_6O_2$ is not formed in this reaction. Although water is present, water does not act as the primary electrophile, nor does it initiate the addition. This is because the electron cloud around oxygen is compact and tightly held, making it difficult to polarise. In contrast, bromine's larger and more diffuse electron cloud makes it far more easily polarised, allowing the bromonium ion to form first. Therefore, a diol is not generated, and bromine must be the first atom to add across the double bond in this mechanism.

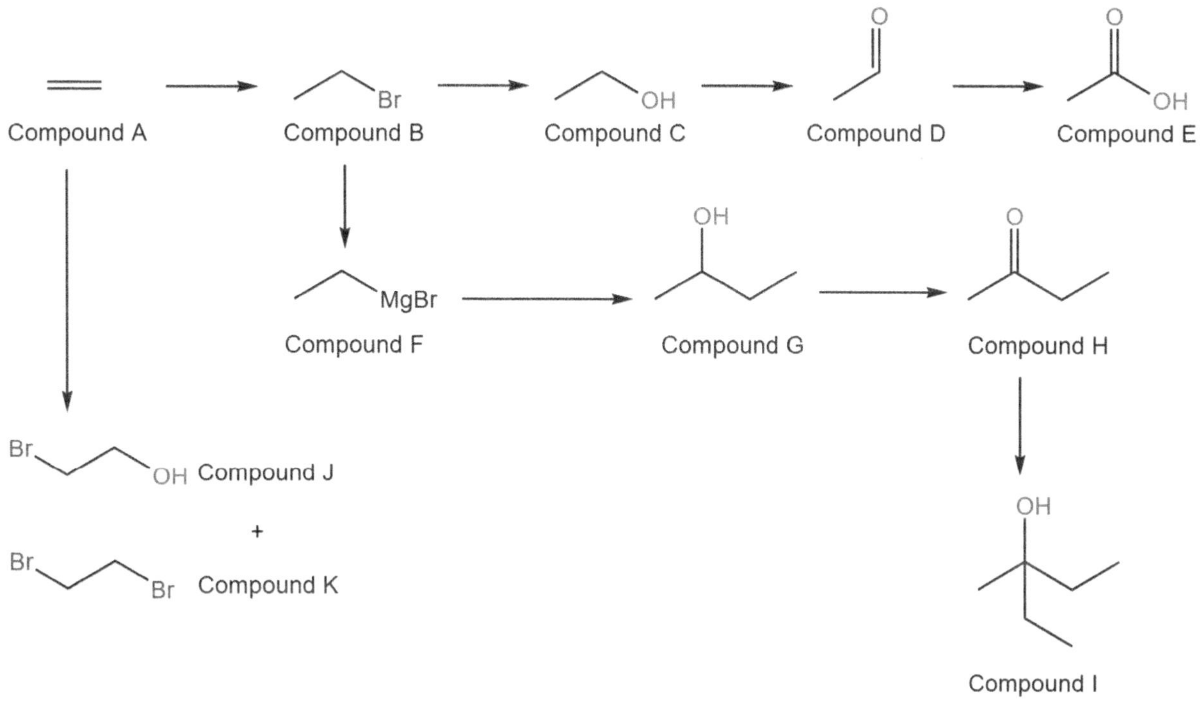

Question 54:

Compound **A** (C_3H_6) reacts with hydrogen bromide in the gaseous phase, forming compound **B** (C_3H_7Br). Upon reaction with aqueous sodium hydroxide, **B** is converted to compound **C** (C_3H_8O), which can undergo oxidation with acidified potassium dichromate, resulting in compound **D** (C_3H_6O). Compound **D** gives a positive result when tested with 2,4-dinitrophenylhydrazine but does not react with Tollen's reagent.

When treated with magnesium in dry ether, compound **B** produces compound **E** (C_3H_7MgBr). This reagent reacts with **D** to yield compound **F** ($C_6H_{14}O$), which remains unreactive when subjected to oxidation with acidified potassium dichromate.

If compound **A** is treated with aqueous bromine, it produces two compounds: **G** (C_3H_7OBr) and **H** ($C_3H_6Br_2$). Notably, $C_3H_8O_2$ is not formed during this reaction. Compound **G** gives a positive result when tested with alkaline aqueous iodine.

(a) Determine the identities and structures of compounds **A** through **H**, providing detailed explanations for all observations and supporting each step with balanced chemical equations.

(b) Explain why $C_3H_8O_2$ is not produced during the reaction involving compound **A** and aqueous bromine.

My thought process:

It is immediately evident that compound **A** is propene. However, when propene undergoes electrophilic addition with hydrogen bromide (HBr), two possible regioisomeric products can form: one via Markovnikov addition, and the other via anti-Markovnikov addition. In structure elucidation questions, it is important not to assume Markovnikov's rule is automatically followed. Instead, one must rely on the experimental clues provided. Assuming a Markovnikov pathway without justification may lead to incorrect conclusions about product identity. Markovnikov's rule is more appropriately applied when planning a synthetic route, not when deducing unknown structures.

Therefore, compound **B** could be either 1-bromopropane, a primary alkyl halide, or 2-bromopropane, a secondary alkyl halide. Upon alkaline hydrolysis, **B** forms compound **C**, an alcohol. Compound **C** is then oxidised to a carbonyl compound **D**, which does not give a positive test with Tollen's reagent. This observation confirms that **D** is a ketone, not an aldehyde. By working backwards, it follows that **C** must be a secondary alcohol, specifically propan-2-ol, and therefore **B** must be 2-bromopropane.

When **B** (2-bromopropane) reacts with magnesium in dry ether, it forms Grignard reagent **E**. This Grignard species carries the nucleophilic character on the secondary carbon of the propyl group. Grignard reagent **E** undergoes nucleophilic addition to compound **D** (propanone) to form alcohol **F**. Since **F** is resistant to oxidation, it must be a tertiary alcohol. However, one does not need to rely on oxidation data alone—drawing the mechanism of nucleophilic addition between the Grignard reagent and ketone is sufficient to determine the exact structure of **F**.

Separately, when propene (**A**) reacts with aqueous bromine (Br_2), the addition begins with the electrophilic attack of Br_2. The bromonium ion forms first, and is then opened by water or another nucleophile. Since bromine is the electrophile, the Br atom must add first, which means a 1,2-diol ($C_3H_8O_2$) cannot be produced directly from this reaction.

Finally, the question reveals that the resulting alcohol **G** gives a positive iodoform test, which indicates that **G** contains the –CH(OH)CH_3 group—a defining feature of certain secondary alcohols. With this critical piece of information, the structure of **G** can be confidently deduced.

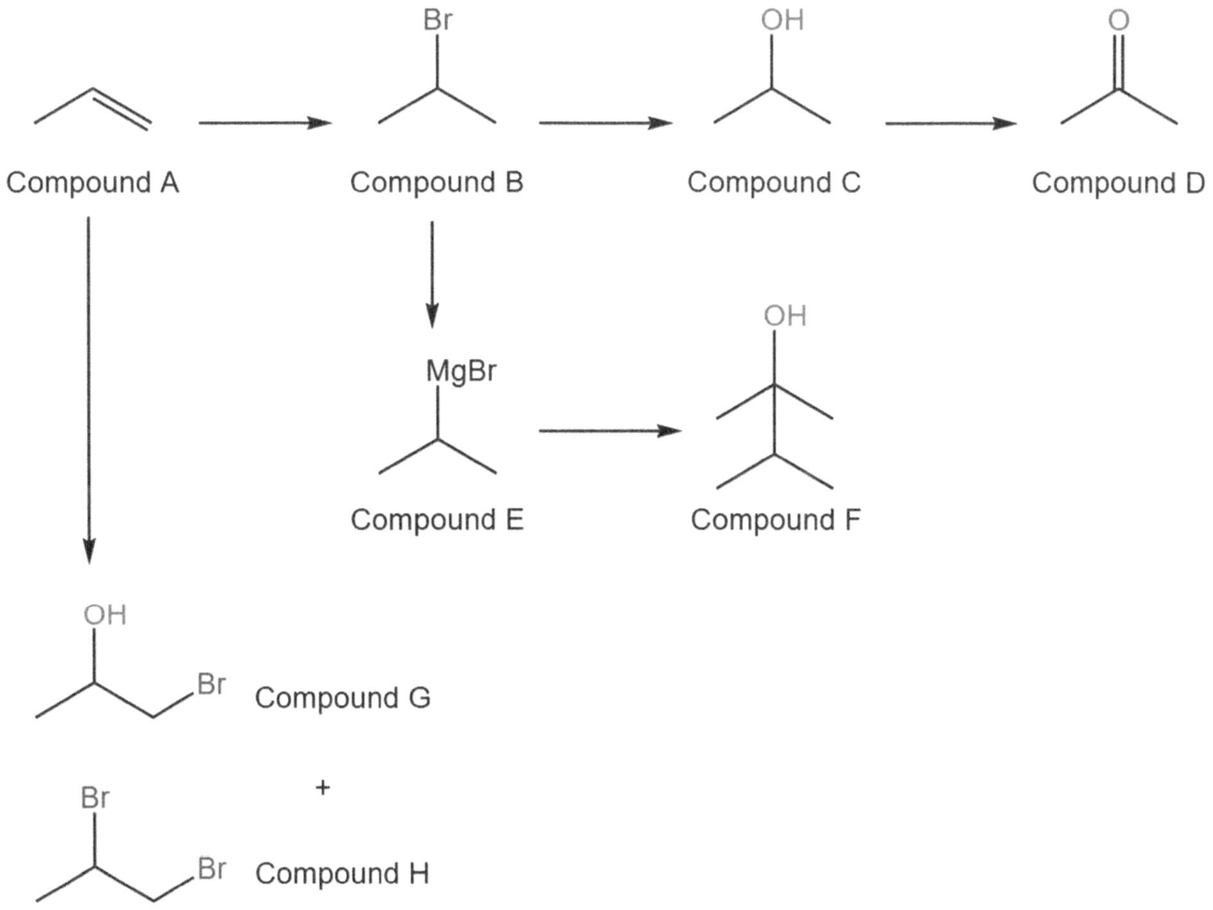

Question 55:

Compound **A** (C_3H_6) reacts with hydrogen bromide in the gaseous phase to produce compound **B** (C_3H_7Br). Upon reaction with aqueous sodium hydroxide, compound **B** is converted into compound **C** (C_3H_8O), which can be oxidized using acidified potassium dichromate to form compound **D** (C_3H_6O). Compound **D** reacts positively with 2,4-dinitrophenylhydrazine but does not produce a reaction with Tollen's reagent.

When compound **B** is treated with magnesium in dry ether, compound **E** (C_3H_7MgBr) is formed. This reagent reacts with compound **D**, producing compound **F** ($C_6H_{14}O$), which remains unreactive towards oxidation with acidified potassium dichromate.

Reacting compound **A** with bromine in the presence of sodium chloride results in the formation of two products: compound **G** (C_3H_6BrCl) and compound **H** ($C_3H_6Br_2$). Subjecting compound **G** to a limited reaction with aqueous sodium hydroxide yields compound **I** as the primary product. Compound **I** does not produce a yellow precipitate when tested with alkaline aqueous iodine solution.

Identify the structural formulas and compositions of compounds **A** through **I**, providing detailed explanations for each observation and supporting conclusions with balanced chemical equations where applicable.

My thought process:

It is evident from the outset that compound **A** is propene. Upon undergoing electrophilic addition, **A** forms compound **B**, though it is initially uncertain whether **B** is 1-bromopropane (a primary alkyl halide) or 2-bromopropane (a secondary alkyl halide). When **B** is treated with aqueous sodium hydroxide, it undergoes nucleophilic substitution to yield alcohol **C**, which could plausibly be either propan-1-ol or propan-2-ol.

Alcohol **C** is then oxidised using acidified potassium dichromate ($K_2Cr_2O_7$) to form compound **D**, which gives a positive result with 2,4-dinitrophenylhydrazine (2,4-DNPH) but a negative result with Tollen's reagent. This confirms that **D** is a ketone, specifically propanone, and not an aldehyde. Working backward, this means that **C** must be propan-2-ol, and consequently, **B** must be 2-bromopropane.

When **B** (2-bromopropane) reacts with magnesium in dry ether, it forms Grignard reagent **E**, in which the nucleophilic carbon resides on the secondary carbon atom of the propyl chain. Grignard reagent **E** undergoes nucleophilic addition to compound **D** (propanone) to yield compound **F**, a tertiary alcohol. Since **F** is resistant to oxidation, its tertiary nature is consistent with the expected outcome.

In another transformation, propene (**A**) reacts with bromine in the presence of sodium chloride in an aqueous medium. In this reaction, the **Br** atom adds first, forming a bromonium ion intermediate, which is then opened by nucleophiles such as water (H_2O), bromide ion (Br^-) from Br_2, or chloride ion (Cl^-) from the dissociation of NaCl. This results in the formation of compound **G**, a dihalogenoalkane containing one bromine and one chlorine atom. However, it is initially unclear whether the product is 1-bromo-2-chloropropane or 2-bromo-1-chloropropane, as both regioisomers are theoretically possible.

The next clue lies in the reaction of **G** with a limited amount of NaOH. Under these conditions, nucleophilic substitution occurs preferentially at the brominated carbon, since the C–Br bond is weaker and more reactive than the C–Cl bond. The resulting alcohol, compound **I**, must be formed in such a way that it does not give a positive iodoform test, ruling out the presence of a –$CH(OH)CH_3$ group.

This leads to the conclusion that **I** is 2-chloropropan-1-ol, which in turn confirms that **G** must be 1-bromo-2-chloropropane.

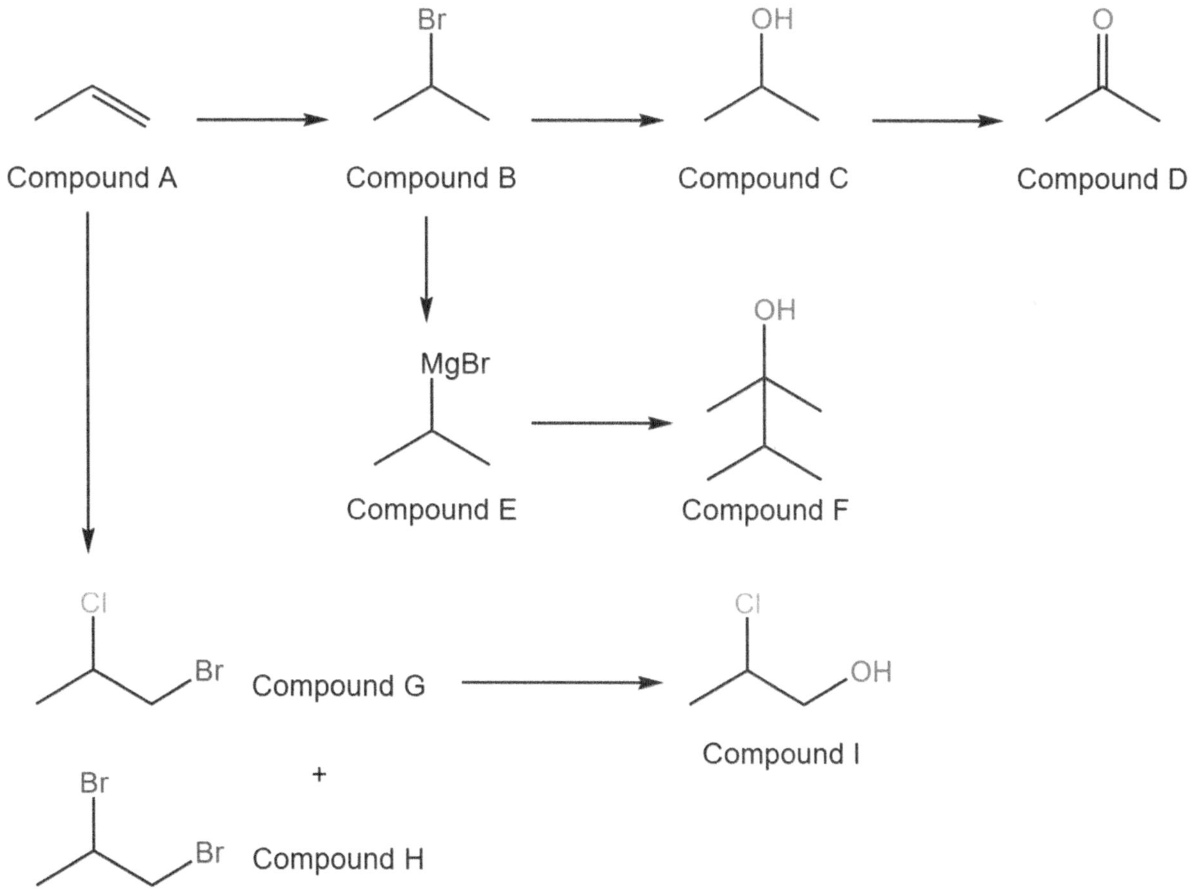

Question 56:

Compound **A** (C_4H_8O) reacts with an alkaline aqueous iodine solution and, upon acidification, forms compound **B** ($C_3H_6O_2$). Compound **A** does not react with Tollen's reagent but produces an orange precipitate with 2,4-dinitrophenylhydrazine.

Reduction of compound **A** using lithium aluminium hydride results in the formation of compound **C** ($C_4H_{10}O$). When treated with phosphorus pentachloride, compound **C** gives compound **D** (C_4H_9Cl). Reaction of **D** with magnesium in dry ether produces **H** (C_4H_9MgCl), which reacts with dry ice and undergoes an aqueous acidic work-up to form compound **I** ($C_5H_{10}O_2$).

Compound **I** reacts with compound **C** in the presence of a catalytic amount of concentrated sulfuric acid, yielding compound **J** ($C_9H_{18}O_2$).

When compound **D** is treated with potassium hydroxide in ethanol, it forms three isomeric products with the molecular formula C_4H_8: a pair of geometric isomers, **E** and **F**, along with compound **G**.

Determine the identities and structural formulas of compounds **A** through **J**, thoroughly explaining all observations. Provide detailed reasoning supported by balanced chemical equations.

My thought process:

Compound **A** has a degree of unsaturation equal to one and contains one oxygen atom, which strongly suggests the presence of a carbonyl functional group. Since **A** gives a positive iodoform test, it must be a methyl ketone. Given that **A** contains four carbon atoms, the only possible methyl ketone with this structure is butanone.

Additional evidence supports this conclusion: **A** reacts positively with 2,4-dinitrophenylhydrazine (2,4-DNPH), confirming the presence of a carbonyl group, and negatively with Tollen's reagent, ruling out an aldehyde. Together, these confirm that **A** is a ketone, specifically butanone.

The iodoform test also provides insight into the identity of compound **B**—the oxidation by-product of **A**. In the test, the $-COCH_3$ group in **A** is cleaved, yielding propanoic acid and a yellow precipitate of triiodomethane (CHI_3).

Reduction of butanone (**A**) with lithium aluminium hydride ($LiAlH_4$) gives compound **C**, butan-2-ol, a secondary alcohol. Treatment of **C** with phosphorus pentachloride (PCl_5) results in nucleophilic substitution, converting the –OH group into a chlorine atom to give compound **D**, 2-chlorobutane.

When compound **D** is reacted with magnesium in dry ether, it forms a Grignard reagent **H**. This Grignard reagent behaves as a butyl nucleophile, with the partial negative charge located on the secondary carbon atom. Upon reaction with dry ice (solid CO_2), **H** attacks the electrophilic carbon in CO_2, forming a carboxylic acid, **I**, after acid work-up.

Compound **I** then undergoes condensation with alcohol **C** (butan-2-ol) to form an ester, compound **J**. This reaction involves nucleophilic acyl substitution, a key transformation in ester synthesis.

In a separate reaction pathway, compound **D**, 2-chlorobutane, can undergo elimination when treated with an alcoholic base, yielding a mixture of alkenes **E**, **F**, and **G**. Since the structure of D is already known, determining the structures of these alkenes—based on Zaitsev's rule and possible E/Z (cis/trans) isomerism—should be straightforward.

🔍 **Tip for students:** Understanding how nucleophiles interact with electrophiles is essential. The mechanism by which Grignard reagents attack carbon dioxide is a classic example of nucleophilic addition, and students are encouraged to refer to *Chapter 4: Arrow Strategy – Core Moves in Organic Mechanisms* for a deeper understanding of this process.

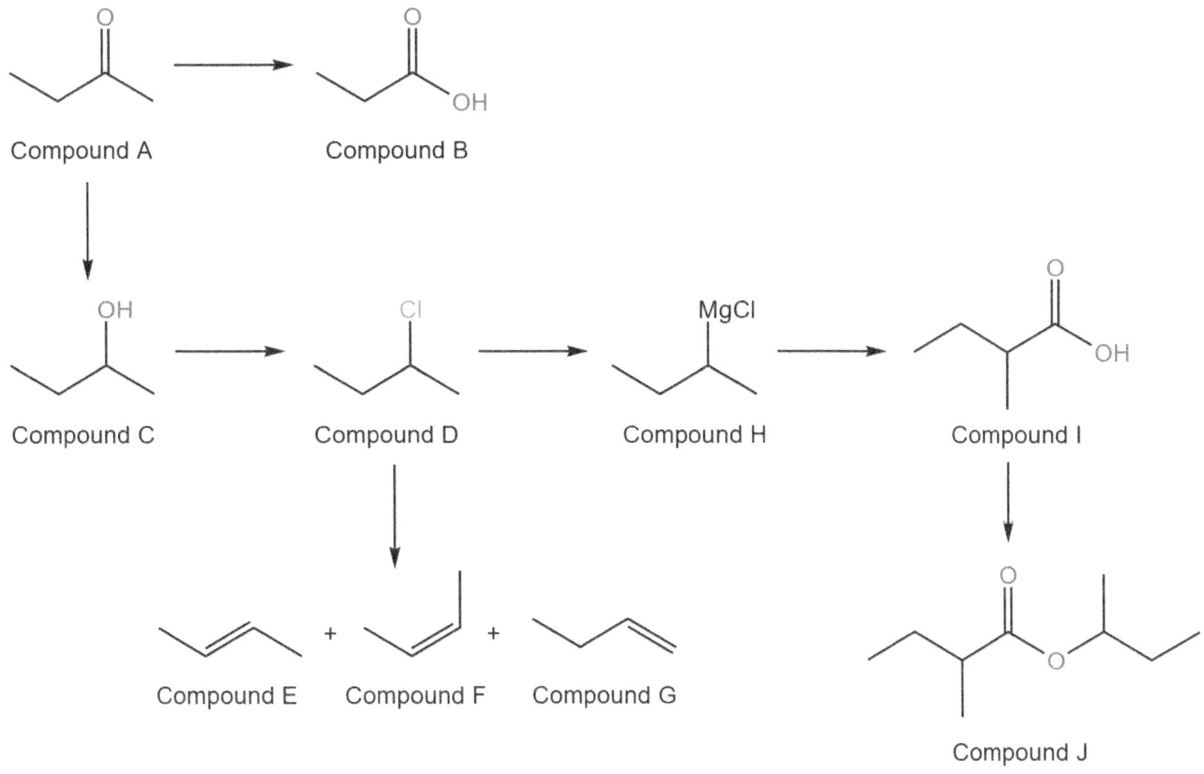

Question 57:

Compound **A** (C_4H_8O) does not undergo a reaction with alkaline aqueous iodine solution. However, it produces an orange precipitate with 2,4-dinitrophenylhydrazine and reacts positively with Tollen's reagent. Reduction of **A** using lithium aluminium hydride yields compound **B** ($C_4H_{10}O$). Treating **B** with phosphorus pentachloride generates compound **C** (C_4H_9Cl).

When compound **C** reacts with magnesium in dry ether, it forms **F** (C_4H_9MgCl). Compound **F** then reacts with dry ice, followed by an acidic work-up, to produce compound **G** ($C_5H_{10}O_2$). Subsequently, **G** reacts with **B** in the presence of a catalytic amount of concentrated sulfuric acid, resulting in the formation of compound **H** ($C_9H_{18}O_2$).

If compound **C** is treated with potassium hydroxide dissolved in ethanol, compound **D** is formed. Oxidizing **D** with potassium permanganate (VII) results in the formation of compound **E**, accompanied by effervescence. Compound **E** also reacts with sodium hydrogen carbonate, producing additional effervescence.

Identify the structural formulas and compositions of compounds **A** through **H**, thoroughly explaining the observations and providing detailed reasoning supported by balanced equations.

My thought process:

Compound **A** has a degree of unsaturation equal to one and contains a single oxygen atom. It gives a positive test with 2,4-dinitrophenylhydrazine (2,4-DNP), indicating the presence of a carbonyl group, and also responds positively to Tollen's reagent, confirming that **A** is an aldehyde. Additionally, since **A** does not give a positive iodoform test, it lacks the $-COCH_3$ moiety and is therefore not a methyl ketone. Given that **A** has four carbon atoms, it could be either butanal (a straight-chain aldehyde) or 2-methylpropanal (a branched-chain aldehyde).

Upon reduction, **A** forms a primary alcohol, compound **B**, which may be either butan-1-ol or 2-methylpropan-1-ol, depending on the structure of **A**. When B is treated with phosphorus pentachloride (PCl_5), it undergoes substitution to form an alkyl chloride, compound **C**, which could be 1-chlorobutane or 1-chloro-2-methylpropane.

The second part of the sequence involves the transformation of compound **C** into a Grignard reagent, **F**, via reaction with magnesium in dry ether. Grignard reagent **F** then reacts with carbon dioxide (CO_2) in a nucleophilic addition reaction to produce a carboxylic acid, compound **G**, after acid work-up. This carboxylic acid **G** subsequently reacts with alcohol **B** in a condensation reaction, forming an ester, compound **H**.

At this point, however, we still lack conclusive evidence to determine whether compound **A** is butanal or 2-methylpropanal, as both candidates remain plausible. The next clue provides a critical distinction: when alkyl chloride **C** undergoes elimination, it yields only one alkene, compound **D**. This is consistent with **C** being a primary alkyl halide, which typically has only one β-hydrogen available for elimination, resulting in a single elimination product.

The oxidation of alkene **D** with acidified potassium manganate(VII) ($KMnO_4$) is accompanied by effervescence, indicating the release of carbon dioxide—a clear sign that **D** is a terminal alkene. Most importantly, the final product of this sequence, compound **E**, is confirmed to be a carboxylic acid because it reacts with sodium hydrogen carbonate, producing effervescence.

By comparing the two potential structures for compound **A** and following through with the transformations, only butanal will yield propanoic acid (**E**) upon oxidative cleavage of a terminal alkene. In contrast, 2-methylpropanal would instead lead to the formation of propanone, a ketone. Thus, butanal is definitively identified as compound **A**.

This question exemplifies how drawing skeletal structures of possible isomers can be a powerful strategy in structure elucidation problems—especially when trying to track transformations step by step.

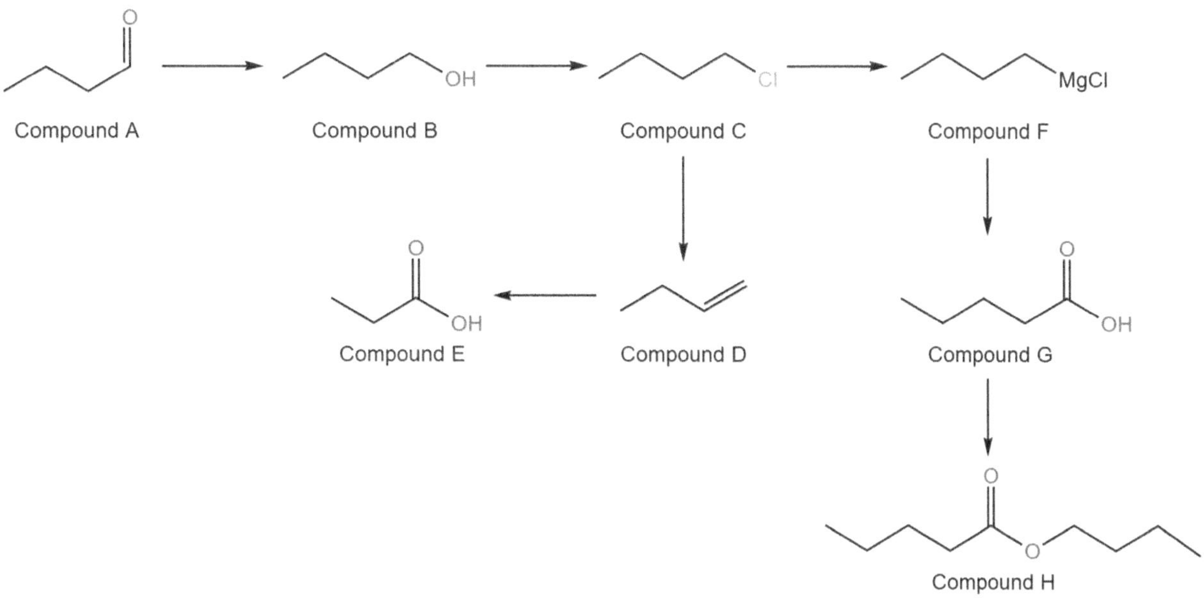

Question 58:

Compound **A** (C_4H_8O) does not undergo a reaction with alkaline aqueous iodine solution. However, it reacts positively with 2,4-dinitrophenylhydrazine and Tollen's reagent. Reduction of **A** using lithium aluminium hydride yields compound **B** ($C_4H_{10}O$). Treating **B** with phosphorus pentachloride results in the formation of compound **C** (C_4H_9Cl).

When compound **C** is reacted with magnesium in dry ether, it produces **F** (C_4H_9MgCl). Compound **F** then reacts with dry ice, followed by acidic work-up, to yield compound **G** ($C_5H_{10}O_2$). Compound **G** reacts with **B** in the presence of a catalytic amount of concentrated sulfuric acid, forming compound **H** ($C_9H_{18}O_2$).

When compound **C** is treated with potassium hydroxide dissolved in ethanol, it forms compound **D**. Oxidation of **D** with potassium permanganate (VII) leads to the formation of compound **E**, with effervescence observed. However, compound **E** does not react with sodium hydrogen carbonate.

Identify the structural formulas and compositions for compounds **A** through **H**, providing detailed explanations of the observations and supporting each conclusion with balanced equations.

My thought process:

Compound **A** has a degree of unsaturation equal to one and contains one oxygen atom, which suggests the presence of a carbonyl group. Its positive reactions with both 2,4-dinitrophenylhydrazine and Tollen's reagent confirm that **A** is an aldehyde. Since **A** consists of four carbon atoms, it could be either butanal (a straight-chain aldehyde) or 2-methylpropanal (a branched isomer).

Upon reduction, **A** forms primary alcohol B, which would be either butan-1-ol or 2-methylpropan-1-ol, depending on the original structure of **A**. Treatment of **B** with PCl_5 yields alkyl chloride **C**—either 1-chlorobutane or 1-chloro-2-methylpropane.

Next, **C** is converted into a Grignard reagent **F** via reaction with magnesium in dry ether. This Grignard reagent undergoes nucleophilic addition to carbon dioxide (CO_2), followed by acidification, to produce carboxylic acid **G**. Compound **G** then reacts with alcohol **B** in a condensation reaction, yielding ester **H**.

At this point, the specific identity of aldehyde **A** is still uncertain, as both possibilities are chemically plausible. The turning point comes with a key observation: **C** undergoes elimination to form only one alkene, **D**. This is expected for a primary alkyl halide, which typically has just one accessible β-hydrogen, allowing only a single elimination product.

The oxidation of alkene **D** with acidified $KMnO_4$ is accompanied by effervescence, indicating the evolution of CO_2, and suggesting that **D** is a terminal alkene. However, the nature of the oxidation product—compound **E**—is crucial. Although compound **E** is obtained after oxidation, it does not react with sodium hydrogen carbonate, confirming that **E** is not a carboxylic acid. The absence of any visible reaction, such as effervescence, strongly suggests that **E** is a ketone—specifically propanone—which does not undergo acid-base reactions with hydrogen carbonate ions.

This subtle but critical observation rules out butanal, which would have led to the formation of propanoic acid, a compound that does react with sodium hydrogen carbonate. Instead, the observed result points to propanone as compound **E**, a ketone that does not effervesce in the presence of sodium hydrogen carbonate.

Since propanone arises from the oxidative cleavage of a terminal alkene only in the case where compound **A** is 2-methylpropanal, we can now confidently conclude that **A** is 2-methylpropanal.

This question closely resembles the logic in Question 57, but hinges entirely on one subtle distinction: whether the final product (compound **E**) reacts with sodium hydrogen carbonate. This small detail plays a decisive role in identifying the correct structure.

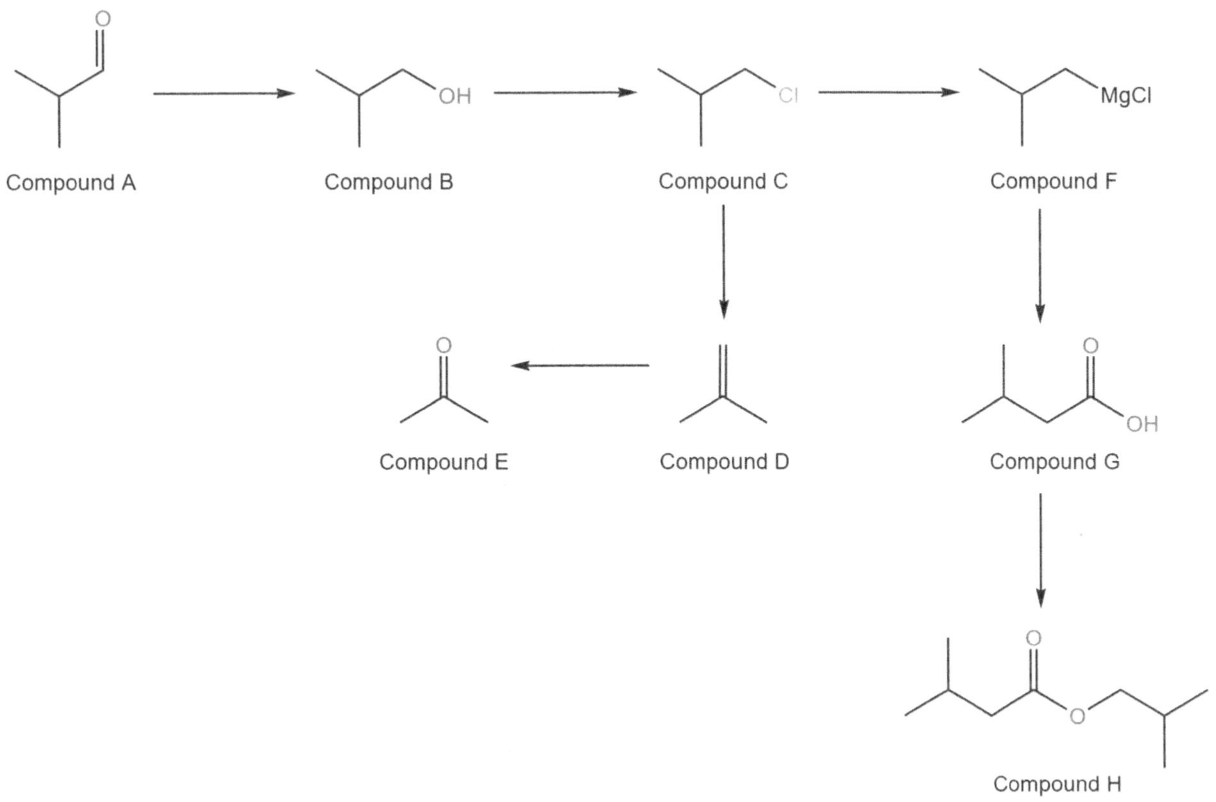

Question 59:

Compound **A** ($C_5H_{10}O$) reacts with alkaline aqueous iodine solution and, upon acidification, forms compound **B** ($C_4H_8O_2$). Compound **A** does not react with Tollen's reagent but produces an orange precipitate with 2,4-dinitrophenylhydrazine.

Reduction of **A** using lithium aluminium hydride gives compound **C** ($C_5H_{12}O$). Treatment of **C** with phosphorus pentachloride results in compound **D** ($C_5H_{11}Cl$). When **D** is reacted with magnesium in dry ether, it yields compound **H** ($C_5H_{11}MgCl$). Compound **H** then reacts with dry ice, followed by aqueous acid work-up, to produce compound **I** ($C_6H_{12}O_2$).

Compound **I** reacts with compound **C** in the presence of a catalytic amount of concentrated sulfuric acid to form compound **J** ($C_{11}H_{22}O_2$).

When compound **D** is treated with potassium hydroxide in ethanol, three isomers of molecular formula C_5H_{10} are formed: a pair of geometric isomers, **E** and **F**, along with compound **G**.

Determine the identities and structural formulas for compounds **A** through **J**, providing a comprehensive explanation for the observations. Include detailed reasoning supported by balanced chemical equations.

My thought process:

Compound **A** has a degree of unsaturation equal to one and contains a single oxygen atom, strongly suggesting the presence of a carbonyl group, as confirmed by its positive reaction with 2,4-dinitrophenylhydrazine (2,4-DNPH). Since **A** does not react with Tollen's reagent, it must be a ketone. Furthermore, **A** gives a positive iodoform test, indicating the presence of a methyl ketone group ($-COCH_3$).

Compound **B** is initially produced as a carboxylate ion, since the iodoform test is carried out under alkaline conditions. After acidification, **B** is converted into the corresponding carboxylic acid, but with the loss of a $-CH_3$ group, which is released as yellow precipitate CHI_3. This explains why **B** contains one carbon atom fewer than **A**.

Given that **A** consists of five carbon atoms, it cannot be an aldehyde. The two most likely structures for a five-carbon methyl ketone are pentan-2-one and 3-methylbutan-2-one.

Reduction of **A** yields a secondary alcohol **C**, which would be either pentan-2-ol or 3-methylbutan-2-ol, depending on the original ketone. Alcohol **C** reacts with phosphorus pentachloride (PCl_5), undergoing nucleophilic substitution to form a secondary alkyl chloride **D**—either 2-chloropentane or 2-chloro-3-methylbutane. **D** then reacts with magnesium in dry ether to form Grignard reagent **H**, which undergoes nucleophilic addition to carbon dioxide (dry ice) to produce a carboxylic acid, compound **I**. This acid subsequently reacts with alcohol **C** in a condensation reaction, forming ester **J**.

Up to this point, the reaction pathway outlines various transformations but does not yet confirm the exact identity of any compound. The key detail is provided when compound **D** undergoes elimination, forming three alkenes (**E**, **F**, and **G**)—two of which are cis-trans isomers. If **D** is 2-chloropentane, it would yield pent-1-ene and cis/trans-pent-2-ene. On the other hand, if **D** were 2-chloro-3-methylbutane, it would form 3-methylbut-1-ene and 2-methylbut-2-ene, which do not show cis-trans isomerism.

Therefore, based on the formation of geometric isomers, compound **D** is confirmed to be 2-chloropentane. Once **D** is identified, the structures of the rest of the compounds can be logically deduced.

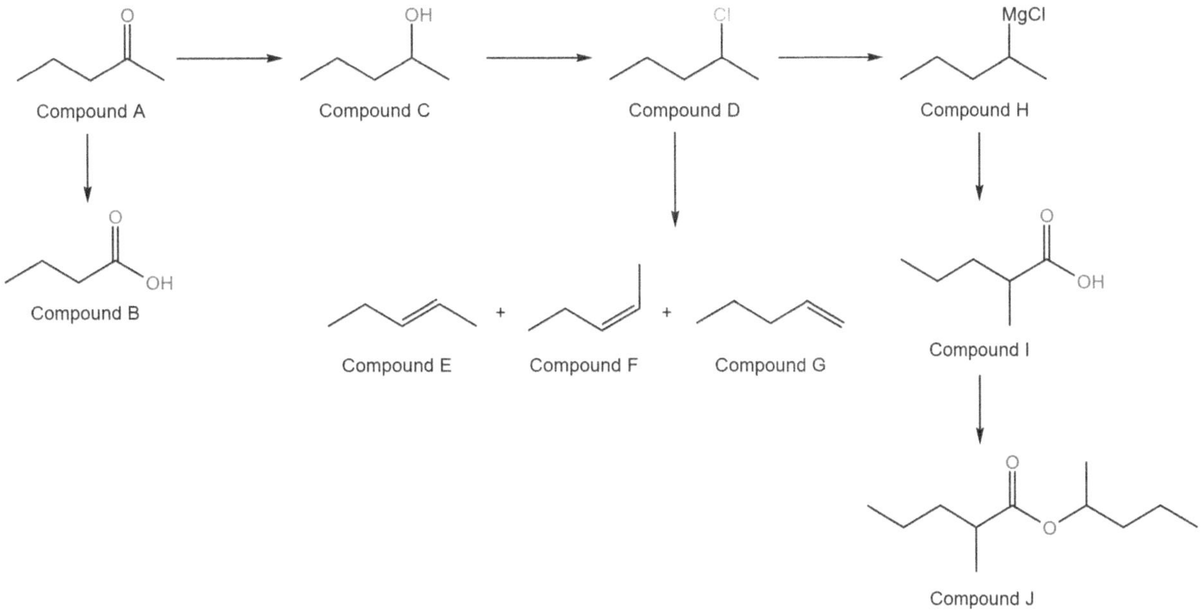

Question 60:

Compound **A**, with the molecular formula $C_5H_{10}O$, reacts with aqueous iodine in the presence of sodium hydroxide. After the reaction mixture is acidified, compound **B** ($C_4H_8O_2$) is obtained. Compound **A** does not react with ammoniacal silver nitrate, but it forms an orange precipitate when treated with 2,4-dinitrophenylhydrazine.

When **A** is treated with lithium aluminium hydride, it is converted into compound **C** ($C_5H_{12}O$). Upon reaction with phosphorus pentachloride, compound **C** gives **D** ($C_5H_{11}Cl$). Compound **D** is then reacted with magnesium in dry ether, forming **G** ($C_5H_{11}MgCl$). This compound reacts with solid carbon dioxide, and after acidic work-up, produces **H** ($C_6H_{12}O_2$).

Compound **H** is reacted with compound **C** in the presence of a catalytic amount of concentrated sulfuric acid, giving compound **I** ($C_{11}H_{22}O_2$).

When compound **D** is treated with potassium hydroxide in ethanol, two isomeric products are formed: **E** and **F**, both with the formula C_5H_{10}. Of these, only **F** produces effervescence when treated with acidified potassium manganate(VII) solution.

Using the information provided, identify the structures of compounds **A** through **I** and explain the observations and transformations involved.

My thought process:

Compound **A** has one degree of unsaturation and contains five carbon atoms. It also gives a positive result with the iodoform test, which suggests the presence of a methyl group adjacent to a carbonyl functional group. Given this information, the only two possible structural isomers that satisfy these conditions are pentan-2-one and 3-methylbutan-2-one. The additional test results stating that **A** does not react with ammoniacal silver nitrate but gives a positive result with 2,4-dinitrophenylhydrazine only confirm what the molecular formula and iodoform test have already strongly implied.

In the course of the iodoform reaction, compound **B** is first formed as a carboxylate salt, as the reaction medium is alkaline. Upon acidification, **B** is converted into the corresponding carboxylic acid, though it now contains one fewer carbon atom due to the elimination of a methyl group as yellow precipitate CHI_3.

Since compound **A** is a ketone, its reduction leads to a secondary alcohol, **C**, which upon treatment with PCl_5 gives the alkyl chloride **D**. **D** reacts with magnesium in dry ether to form the Grignard reagent **G**, which then adds to carbon dioxide, producing the carboxylic acid **H** after acidic work-up. Carboxylic acid **H** acid undergoes condensation with the secondary alcohol **C** to form ester **I**.

While these transformations demonstrate a clear synthetic route, they do not yet distinguish between the two structural options for compound **A**. The decisive clue is that compound **D** undergoes elimination to yield two alkenes, **E** and **F**, and that **F** gives effervescence with acidified potassium manganate(VII)—a sign of oxidative cleavage at a terminal alkene.

Given this observation, we can work backward and deduce that 3-methylbutan-2-one must be the correct structure for compound **A**. This is the only candidate whose downstream intermediates are consistent with the given chemical behaviour and products observed.

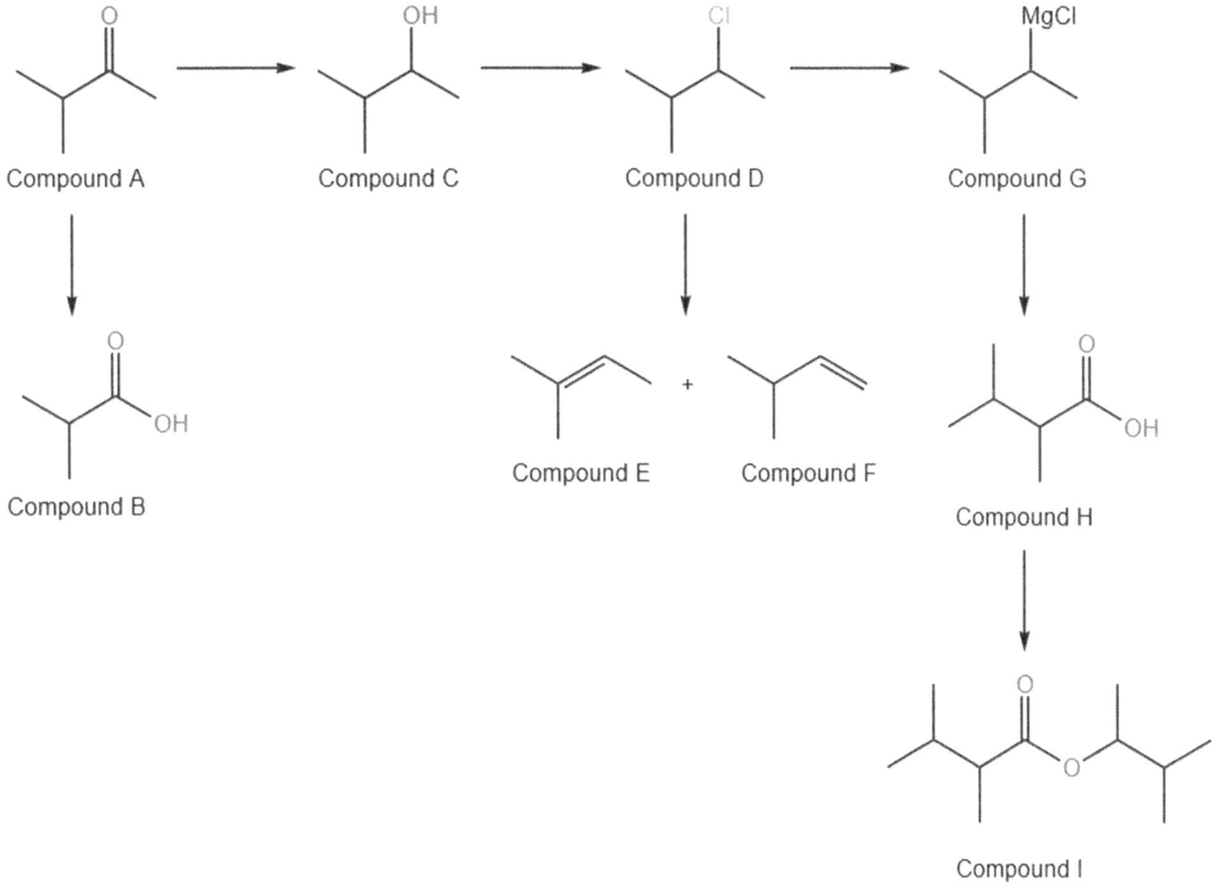

15 Answers to Chapter 12 — C, H, O, X, and N

As before, follow the logical progression shown in each flowchart to verify your answers. Don't forget to test yourself by filling in the missing details for each transformation.

Question 61:

Compound **A** (C_4H_9Cl) undergoes a reaction with aqueous sodium hydroxide to produce **B**. When reacted with ammonia, **A** yields **D**. Additionally, treating **A** with potassium cyanide dissolved in ethanol forms **E** (C_5H_9N).

Compound **E** reacts with lithium aluminium hydride in dry ether, resulting in the formation of **F**. When **E** undergoes hydrolysis with aqueous sodium hydroxide followed by treatment with dilute hydrochloric acid, it produces **G** ($C_5H_{10}O_2$). Further reaction of **G** with phosphorus(V) chloride leads to the formation of **H**.

Treating **A** with potassium hydroxide in ethanol leads to the formation of **I**. When compounds **B** and **I** are oxidized using potassium manganate(VII), they yield **C** and **J**, respectively. Compounds **C** ($C_4H_8O_2$) and **J** ($C_3H_6O_2$) both react with sodium hydrogen carbonate to produce effervescence.

Your task is to determine the identities of compounds **A** through **J**, providing detailed reasoning for each step and including the relevant chemical equations.

My thought process:

Compound **A** has a degree of unsaturation of zero and contains one chlorine atom, indicating that it is a saturated alkyl chloride. At this stage, we don't know whether **A** is a straight-chain or branched halide, nor whether it is primary, secondary, or tertiary. The transformation of **A** into compounds **B**, **D**, and **E** occurs via nucleophilic substitution.

Compound **E** is a nitrile that, when reduced with $LiAlH_4$, gives a primary amine **F** only.

🔍 **Tip for students:** Reducing a nitrile always yields a primary amine—not a secondary or tertiary amine).

In addition, **E** undergoes acidic hydrolysis to form compound **G**, which is a carboxylic acid; **G** then reacts with PCl_5 through nucleophilic acyl substitution to produce an acyl chloride. Meanwhile, **A** also undergoes dehydrohalogenation to yield alkene **I** (maintaining a four-carbon framework).

We know that compound **B** is an alcohol, although at first its exact nature as primary, secondary, or tertiary remains uncertain because that depends on the classification of **A**. However, when **B** is oxidized with $KMnO_4$, it forms compound **C** ($C_4H_8O_2$) which has a degree of unsaturation of one and contains two oxygen atoms; this suggests that **C** is a carboxylic acid. Similarly, compound **J**, which also shows one degree of unsaturation and two oxygen atoms, is deduced to be a carboxylic acid as well—its identity as a carboxylic acid is confirmed by the effervescence produced when it reacts with sodium hydrogen carbonate. Given that **J** is a three-carbon carboxylic acid, it must be propanoic acid.

Because **J** (with three carbons) is produced from the oxidative cleavage of alkene **I** (which has four carbons), the loss of one carbon atom indicates that **I** is a terminal alkene; hence, **I** is but-1-ene. Since **C** is obtained by oxidizing alcohol **B** with $KMnO_4$ and is a carboxylic acid, we deduce that **B** is a primary alcohol. Therefore, **A** must be a primary alkyl halide.

The only remaining issue is whether **A** has a straight-chain or branched structure. For a four-carbon primary alkyl halide, there are only two options—1-chlorobutane or 1-chloro-2-methylpropane. However, because the structure of **I** is determined to be but-1-ene (obtained by elimination of HCl from **A**), the carbon skeleton of **A** must match that of **I**. Thus, **A** can only be 1-chlorobutane. Once **A** is identified as 1-chlorobutane, the structures of compounds **B** through **J** follow naturally.

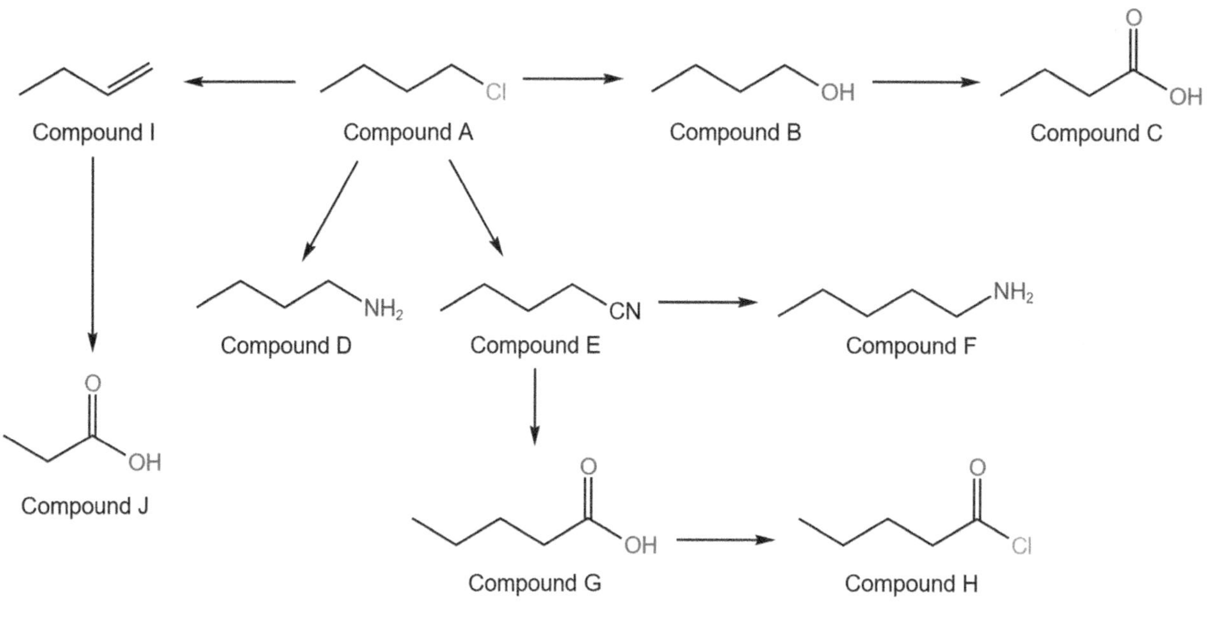

Question 62:

Compound **A** (C_4H_9Cl) reacts with aqueous sodium hydroxide, yielding **B** ($C_4H_{10}O$). When treated with ammonia, **A** forms **D** ($C_4H_{11}N$). Reaction of **A** with potassium cyanide in ethanol produces **E** (C_5H_9N).

Compound **E** reacts with lithium aluminium hydride in dry ether, resulting in **F** ($C_5H_{13}N$). Additionally, hydrolysis of **E** with aqueous sodium hydroxide followed by treatment with dilute hydrochloric acid leads to **G** ($C_5H_{10}O_2$). Reacting **G** with phosphorus(V) chloride produces **H** (C_5H_9OCl).

If **A** undergoes treatment with potassium hydroxide in ethanol, three isomeric compounds—**J**, **K**, and **I** (all with the molecular formula C_4H_8)—are obtained. Oxidation of **J** and **K** with potassium manganate(VII) yields **M** ($C_2H_4O_2$), while similar treatment of **I** results in **L** ($C_3H_6O_2$). When compound **B** is oxidized under the same conditions, it produces **C** (C_4H_8O).

Compound **C** does not react with ammoniacal silver nitrate but forms an orange precipitate, **N**, upon reacting with 2,4-dinitrophenylhydrazine. Reduction of **N** using tin and excess concentrated hydrochloric acid followed by treatment with aqueous sodium hydroxide yields **O** ($C_{10}H_{16}N_4$).

Determine the identities of compounds **A** through **O**, explaining your reasoning and including relevant chemical equations to justify each step.

My thought process:

Compound **A** possesses a degree of unsaturation of zero and contains one chlorine atom, signifying that it is a saturated alkyl halide. At this early stage, we cannot yet determine whether it is a straight-chain or branched halide, or whether it is primary, secondary, or tertiary. The conversion of **A** into alcohol **B**, amine **D**, and nitrile **E** via nucleophilic substitution provides further clues about its structure.

Nitrile **E** is reduced to form primary amine **F**, and its acidic hydrolysis produces carboxylic acid **G**, which subsequently reacts with PCl_5 in a nucleophilic acyl substitution to yield an acyl chloride. Additionally, **A** undergoes dehydrohalogenation to produce a set of alkenes—**J**, **K**, and **I**.

🔍 **Tip for students:** If oxidative cleavage of two isomeric alkenes results in only one product, then those alkenes must be **symmetrical**. If they also contain two different groups on each carbon of the C=C bond, then they are **geometrical isomers**.

Since **J** and **K** yield a single product, **M**, they must be cis/trans isomers and symmetrical. **M**, with a degree of unsaturation of one and two oxygen atoms, corresponds to a two-carbon carboxylic acid, such as ethanoic acid; hence, **J** and **K** can be identified as cis/trans but-2-ene.

Compound **C**, formed in the presence of $KMnO_4$, exhibits a degree of unsaturation of one and contains one oxygen atom, which indicates that it is a four-carbon ketone—for example, butanone. Once **C** is recognized as butanone, the structures of the other compounds follow naturally. Specifically, if **C** is butanone, then alcohol **B** must be butan-2-ol, implying that **A** is 2-chloro-butane.

In this problem, the key is to first determine the structure of compound **C**, after which the structures of **A**, **B**, and the remaining compounds become clear.

Note for students: I intentionally did not mention at the outset that compound **C** is the "weak link" in the series; this insight emerged only upon working through the entire problem. My aim here is to share my thinking process rather than simply presenting the final answer.

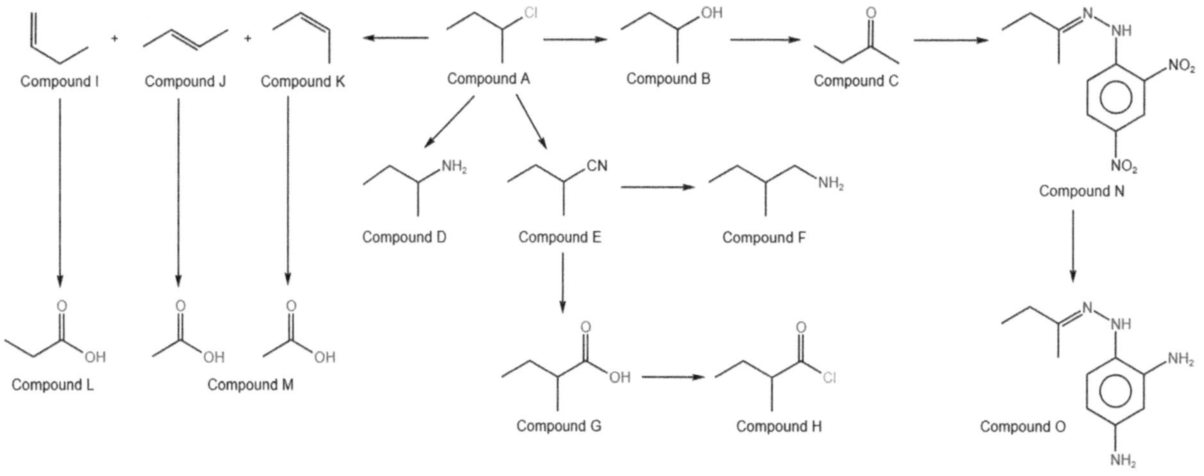

Question 63:

Compound **A** (C_4H_9Cl) undergoes a reaction with aqueous sodium hydroxide to produce **B** ($C_4H_{10}O$). When **A** reacts with ammonia, it forms **C** ($C_4H_{11}N$). Additionally, treating **A** with potassium cyanide in ethanol results in the formation of **D** (C_5H_9N).

Compound **D** reacts with lithium aluminium hydride in dry ether to yield **E** ($C_5H_{13}N$). Hydrolysis of **D**, first with aqueous sodium hydroxide followed by treatment with dilute hydrochloric acid, leads to **F** ($C_5H_{10}O_2$). Further reaction of **F** with phosphorus(V) chloride generates **G** (C_5H_9OCl).

When **A** is treated with potassium hydroxide in ethanol, it forms **H**, a compound with the molecular formula C_4H_8. Oxidation of **H** using potassium manganate(VII) results in the formation of **I** (C_3H_6O), with effervescence observed during the reaction. On the other hand, if compound **B** is subjected to the same oxidative conditions, no new product is formed.

Your task is to identify compounds **A** through **I**. For each step, provide detailed reasoning and include the relevant chemical equations to justify your conclusions.

My thought process:

Compound **A** exhibits zero degrees of unsaturation and contains one chlorine atom, which indicates that it is a saturated alkyl chloride. The formation of alcohol **B**, amine **C**, and nitrile **D** occurs via the typical nucleophilic substitution reactions that alkyl halides undergo.

Nitrile **D** is reduced to produce primary amine **E**, and it also undergoes acidic hydrolysis to yield carboxylic acid **F**, which subsequently reacts with PCl_5 via nucleophilic acyl substitution to generate acyl chloride **G**.

Although we know that **A** is a saturated alkyl chloride, its structure is not yet clear in terms of whether it is a straight-chain or branched alkyl halide, or whether it is primary, secondary, or tertiary. However, upon dehydrohalogenation, **A** produces only one alkene, **H**. This observation is a key clue in narrowing down the structure of **A**. When alkene **H** is subjected to strong oxidation with $KMnO_4$, it gives compound **I**, which is identified as a ketone. Since **I** is a three-carbon ketone, it must be propanone. The fact that there is a one-carbon difference between alkene **H** and ketone **I** indicates that **H** is a terminal alkene, and therefore **H** can only be 2-methylpropene.

Another important clue is that alcohol **B** does not undergo oxidation with $KMnO_4$, which suggests that **B** is a tertiary alcohol. If **B** is tertiary, then **A** must be a tertiary alkyl chloride. Considering that **A** is a four-carbon tertiary alkyl chloride, there is only one possibility: **A** must be 2-chloro-2-methylpropane. Once **A** has been identified, the structures of compounds **B** through **I** can be determined naturally.

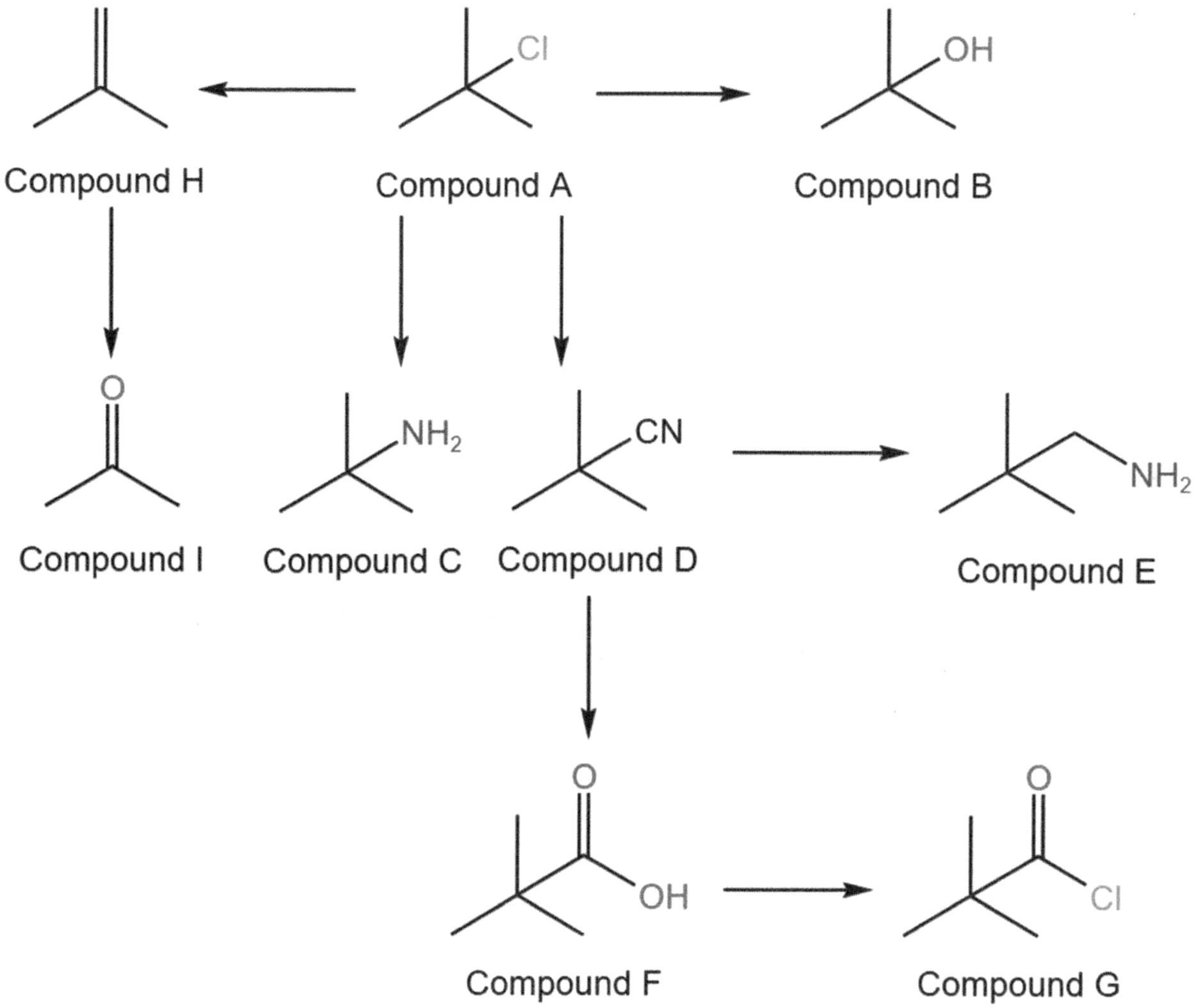
Cl
OH
Compound H
Compound A
Compound B
O
NH2
CN
NH2
Compound I
Compound C
Compound D
Compound E
O
OH
O
Cl
Compound F
Compound G

Question 64:

Compound **A** (C_4H_9Cl) reacts with aqueous sodium hydroxide, leading to the formation of **B** ($C_4H_{10}O$). Treating **A** with ammonia produces **D** ($C_4H_{11}N$). Additionally, when **A** is exposed to potassium cyanide dissolved in ethanol, it generates **E** (C_5H_9N).

Compound **E** undergoes reduction with lithium aluminium hydride in dry ether, resulting in **F** ($C_5H_{13}N$). Hydrolysis of **E**, first using aqueous sodium hydroxide followed by dilute hydrochloric acid, gives **G** ($C_5H_{10}O_2$). Subsequent reaction of **G** with phosphorus(V) chloride produces **H** (C_5H_9OCl).

If **A** is treated with potassium hydroxide in ethanol, it forms **I**, a compound with the molecular formula C_4H_8. Oxidation of **I** using potassium manganate(VII) results in **J** (C_3H_6O), accompanied by the observation of effervescence. Under similar oxidative conditions, compound **B** produces a new product, **C**, that can react with sodium metal to generate effervescence and a characteristic 'pop' sound, extinguishing a lighted splint.

Your task is to identify compounds **A** through **J**, providing detailed reasoning for your conclusions and supporting them with relevant chemical equations.

My thought process:

Compound **A** has no degrees of unsaturation and contains one chlorine atom, which indicates it is a saturated alkyl halide. The formation of alcohol **B**, amine **D**, and nitrile **E** proceeds via the typical nucleophilic substitution reactions that alkyl halides undergo.

Nitrile **E** is reduced to form primary amine **F**. Additionally, when nitrile **E** is subjected to acidic hydrolysis, it yields carboxylic acid **G**; this acid then reacts with PCl_5 through nucleophilic acyl substitution to form acyl chloride **H**.

At this point, it remains unclear whether **A** is a straight-chain or branched alkyl halide, or whether it is primary, secondary, or tertiary. However, the dehydrohalogenation of **A** produces only one alkene, **I**. The oxidative cleavage of alkene **I** gives compound **J**, which is a three-carbon ketone (for example, propanone). Since there is a one-carbon difference between four-carbon alkene **I** and the three-carbon ketone (**J**), alkene **I** must be 2-methylpropene.

Although we know that **B** is an alcohol, it is initially uncertain whether it is primary, secondary, or tertiary. However, when alcohol **B** is subjected to strong oxidation with $KMnO_4$, the product must retain a hydroxyl group so that it can react with sodium metal to produce hydrogen gas. This observation indicates that **B** is a primary alcohol, because a primary alcohol will be oxidized to a carboxylic acid that can react with sodium metal, whereas a tertiary alcohol will not oxidize and a secondary alcohol would yield a ketone.

If we know the structure of alkene **I** and accept that **A** is a primary alkyl chloride, then the addition of H and Cl must occur in such a way to yield a primary alkyl chloride **A**. Consequently, **A** can only be 1-chloro-2-methylpropane. Once **A** is identified, the structures for compounds **B** through **J** follow naturally.

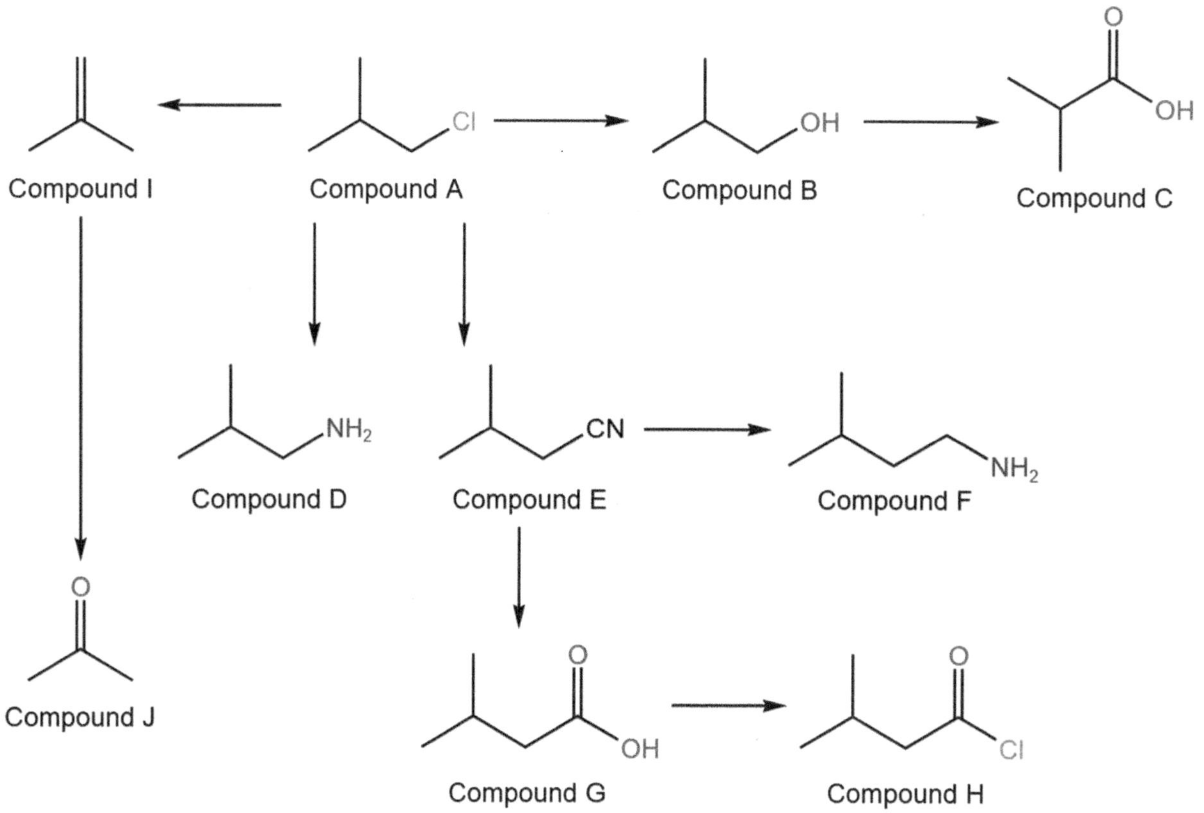

Question 65:

Compound **A** ($C_9H_8O_2$), a solid, effervesces upon contact with sodium bicarbonate solution, releasing carbon dioxide gas. When treated with bromine water, **A** reacts immediately, resulting in decolourisation and the formation of **B** ($C_9H_8O_2Br_2$). Reacting **B** with aqueous sodium hydroxide produces the sodium salt of **C** ($C_9H_{10}O_4$).

Through oxidation using acidified potassium dichromate, **C** is transformed into **D** ($C_9H_6O_4$). Compound **D** shows a positive response to 2,4-dinitrophenylhydrazine but remains unreactive with ammoniacal silver nitrate solution. Upon treatment with phosphorus pentachloride, **D** releases steamy fumes, forming **E** ($C_9H_5O_3Cl$).

Further reaction between **E** and ammonia produces **F** ($C_9H_7O_3N$). All compounds from **A** through **F** undergo oxidation with hot acidified aqueous potassium manganate (VII), resulting in **G** ($C_7H_6O_2$).

Your objective is to identify compounds **A** through **G**. Explain the observations for each step and support your answers with relevant chemical equations.

My thought process:

Compound **A** has a degree of unsaturation of 6, which immediately suggests the presence of a phenyl ring (since a DU of 4 or more typically signals a benzene ring). Its reaction with sodium bicarbonate, which releases carbon dioxide, indicates that **A** contains a carboxyl group. In a molecule with a phenyl group (4 DU) and a carboxyl group (1 DU), the total would be 5; therefore, the extra degree of unsaturation suggests that an additional unsaturation, such as an alkene, is present in **A**. The fact that **A** also reacts with bromine confirms the presence of a C=C bond.

Thus, we can conclude that **A** comprises a phenyl ring, a carboxyl group, and an alkene, adding up to a total of 6 degrees of unsaturation. Next, an electrophilic addition reaction with bromine occurs across the C=C bond in **A**, leading to the formation of compound **B**, which is characterized as a dihalogenoalkane. Following this, **B** undergoes nucleophilic substitution to yield compound **C**, a diol.

When compound **C** (which initially has a degree of unsaturation of 5) is oxidized, compound **D** is formed with a degree of unsaturation of 7. The increase by 2 degrees is due to the oxidation of both secondary alcohol groups (present in the diol) into ketone groups—a change confirmed by the positive 2,4-dinitrophenylhydrazine test. Compound **D**, which still retains a carboxyl group, then undergoes nucleophilic acyl substitution with phosphorus pentachloride to form compound **E**, an acyl chloride.

Since **E** contains an acyl chloride group, it reacts with ammonia to form compound **F**, an amide. Lastly, when a compound with the molecular formula $C_7H_6O_2$ is observed, our first instinct is to identify it as benzoic acid. Given that **A** is oxidized with $KMnO_4$ to produce benzoic acid, this indicates that there is only one substituent on the phenyl ring, and that this substituent consists of both an alkene and a carboxyl group. This insight is sufficient for us to deduce the structure of **A**, and once **A** is established, the structures of compounds **B** through **G** follow naturally.

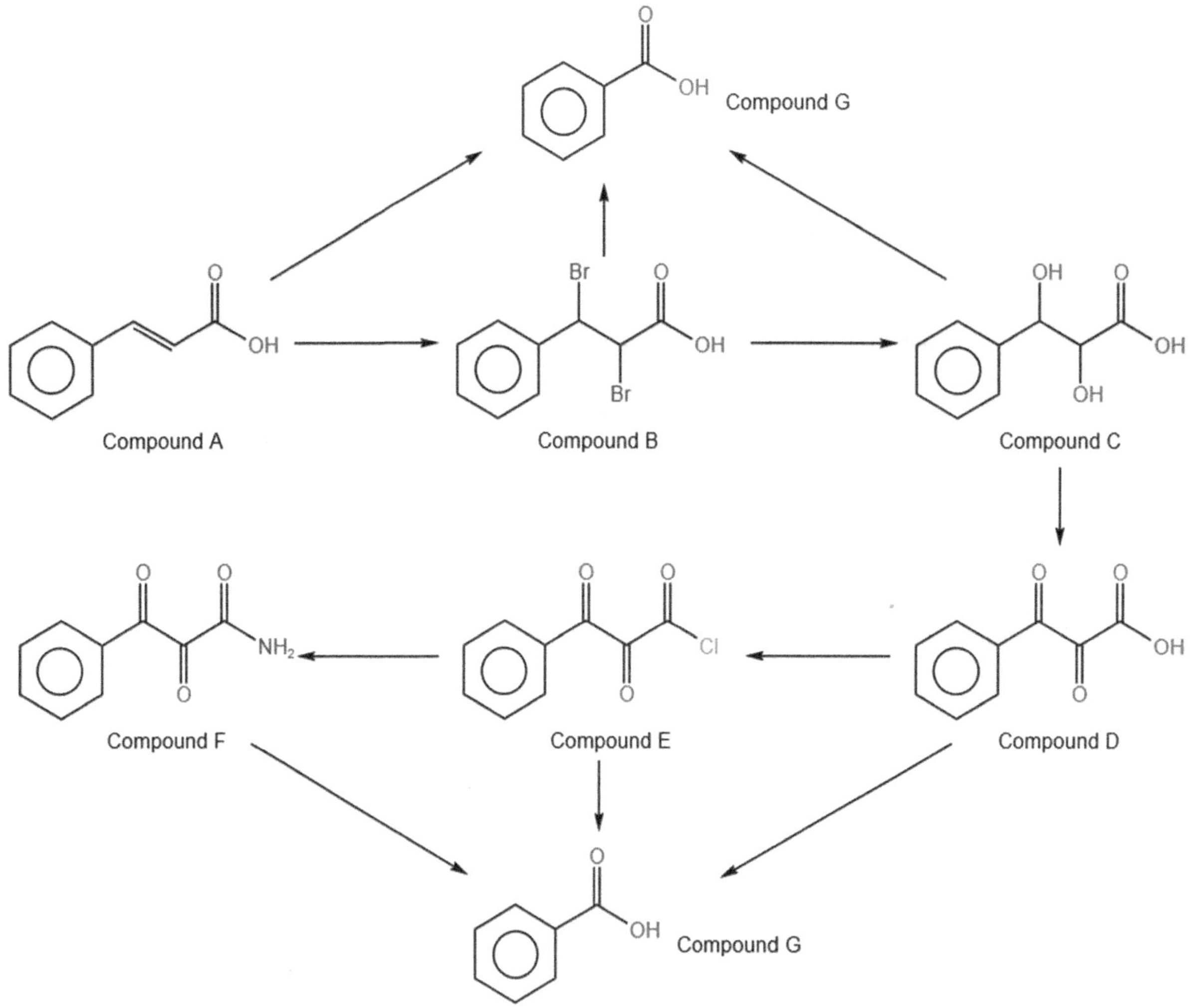

Question 66:

Compound **A** ($C_{10}H_{12}$) reacts with acidified potassium manganate(VII), causing decolourisation of the solution and resulting in the formation of **B** ($C_8H_6O_4$). When **B** is treated with lithium aluminium hydride in dry ether, it produces **C** ($C_8H_{10}O_2$). Upon treatment with sodium metal followed by chloromethane, **C** is converted into **E** ($C_{10}H_{14}O_2$).

When compound **C** is reacted with phosphorus trichloride, it releases steamy fumes and forms **D** ($C_8H_8Cl_2$). Further treatment of **D** with an excess of ammonia leads to the formation of **F** ($C_8H_{12}N_2$). However, exposing **F** to tin and concentrated hydrochloric acid in excess does not result in the formation of any new product.

Your task is to determine the identities of compounds **A** through **F**, providing detailed explanations for all observations and supporting your conclusions with relevant chemical equations.

My thought process:

Compound **A** has a degree of unsaturation (DU) of 5, which immediately hints at the presence of a benzene ring because a benzene ring contributes 4 units of unsaturation. With the molecular formula $C_8H_6O_4$, it becomes clear that the benzene ring is substituted with two carboxyl groups. In essence, compound **A** consists of a phenyl ring (providing DU = 4) along with an additional cyclic element (contributing the extra DU = 1).

Once you determine the structure of **A**, the formation of compounds **B** through **F** follows naturally. First, compound **B** is reduced, specifically converting its two carboxyl groups into hydroxyl groups to yield a diol (compound **C**). When compound **C** is treated with sodium metal, the hydroxyl groups are deprotonated to form alkoxide ions, which then act as nucleophiles and attack chloromethane, leading to the formation of an ether (compound **E**) via the Williamson ether synthesis mechanism.

Alternatively, the same diol (**C**) can react with phosphorus trichloride (PCl_3) in a nucleophilic substitution to form a halogenoalkane (compound **D**). This intermediate then reacts with ammonia in another nucleophilic substitution, resulting in the formation of an amine (compound **F**). The fact that compound **F** is not reduced by tin and concentrated HCl indicates that it is an aliphatic amine, rather than an aromatic one like phenylamine.

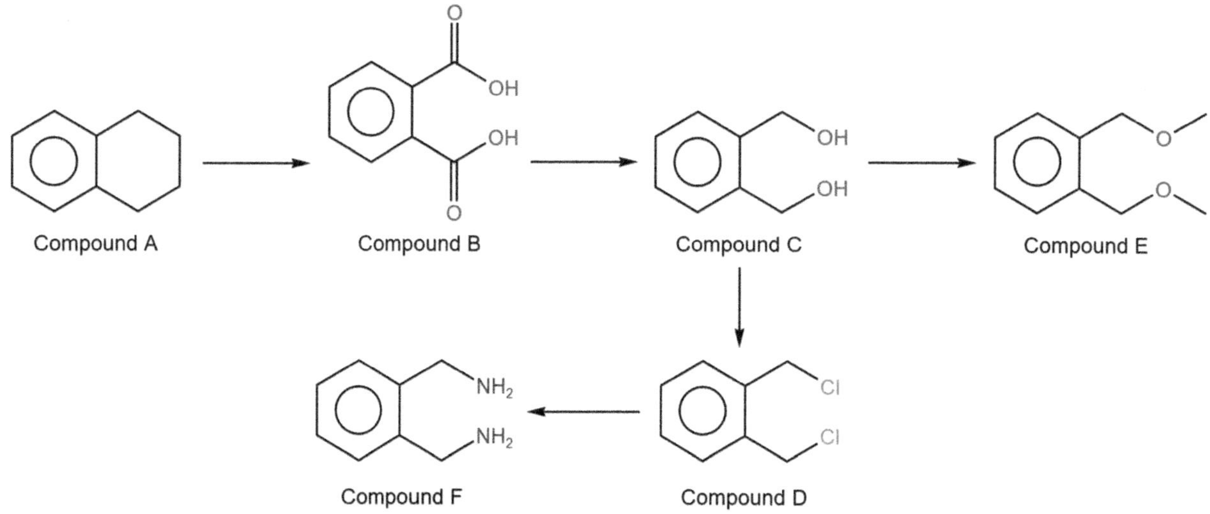
OH
OH
OH
OH
NH2
NH2
Cl
Cl
Compound A
Compound B
Compound C
Compound E
Compound F
Compound D

Question 67:

Compound **A** ($C_{10}H_8$) reacts with acidified potassium manganate(VII), causing decolourisation of the solution and yielding **B** ($C_8H_6O_4$). When **B** is subjected to heating with soda-lime—a mixture of calcium oxide and sodium hydroxide—it undergoes decarboxylation, producing **C** (C_6H_6).

Compound **C** reacts with chloromethane in the presence of anhydrous iron(III) chloride, resulting in the formation of **D** (C_7H_8). Like **A**, **D** decolourises acidified potassium manganate(VII) solution, forming **E** ($C_7H_6O_2$). Upon treatment with hot concentrated sulfuric acid and nitric acid, **E** undergoes nitration, producing **F** ($C_7H_5NO_3$).

When **F** is reacted with tin and concentrated hydrochloric acid, it is converted to a salt of **G** (C_7H_7N). Compound **G** reacts rapidly with aqueous bromine, decolourising the solution and producing **H** ($C_7H_4O_2NBr_3$).

Identify and explain the observations for compounds **A** through **H**, detailing each transformation step and supporting your conclusions with relevant chemical equations.

My thought process:

Compound **A** has a degree of unsaturation (DU) of 7, which immediately indicates the presence of a benzene ring because a benzene ring alone accounts for 4 DU. The molecular formula $C_8H_6O_4$ further suggests that this benzene ring is substituted with two carboxyl groups. From these clues, I deduced that compound **A** must include a phenyl ring (contributing DU = 4) along with an additional cyclic structure (contributing at least DU = 1) adjacent to it, with the remaining 2 DU most likely coming from two alkene groups within that cyclic ring. Typically, such a cyclic ring has either five or six members to be thermodynamically favorable.

It is important to note that the later observations from the question—where compound **C** is described simply as a benzene ring undergoing electrophilic substitution with CH_3Cl to give methylbenzene **D**, which is then oxidized to form benzoic acid E ($C_7H_6O_2$) and further transformed through nitration, reduction, and subsequent bromination—do not contribute to solving the structure of compound **A**. Instead, these steps serve to confirm the substitution pattern on the benzene ring. The key to predicting the structure of **A** lies solely in understanding the degree of unsaturation and the given molecular formula. Thus, recognizing that DU is 7 was crucial for deducing the structure of **A** independently of the subsequent observations.

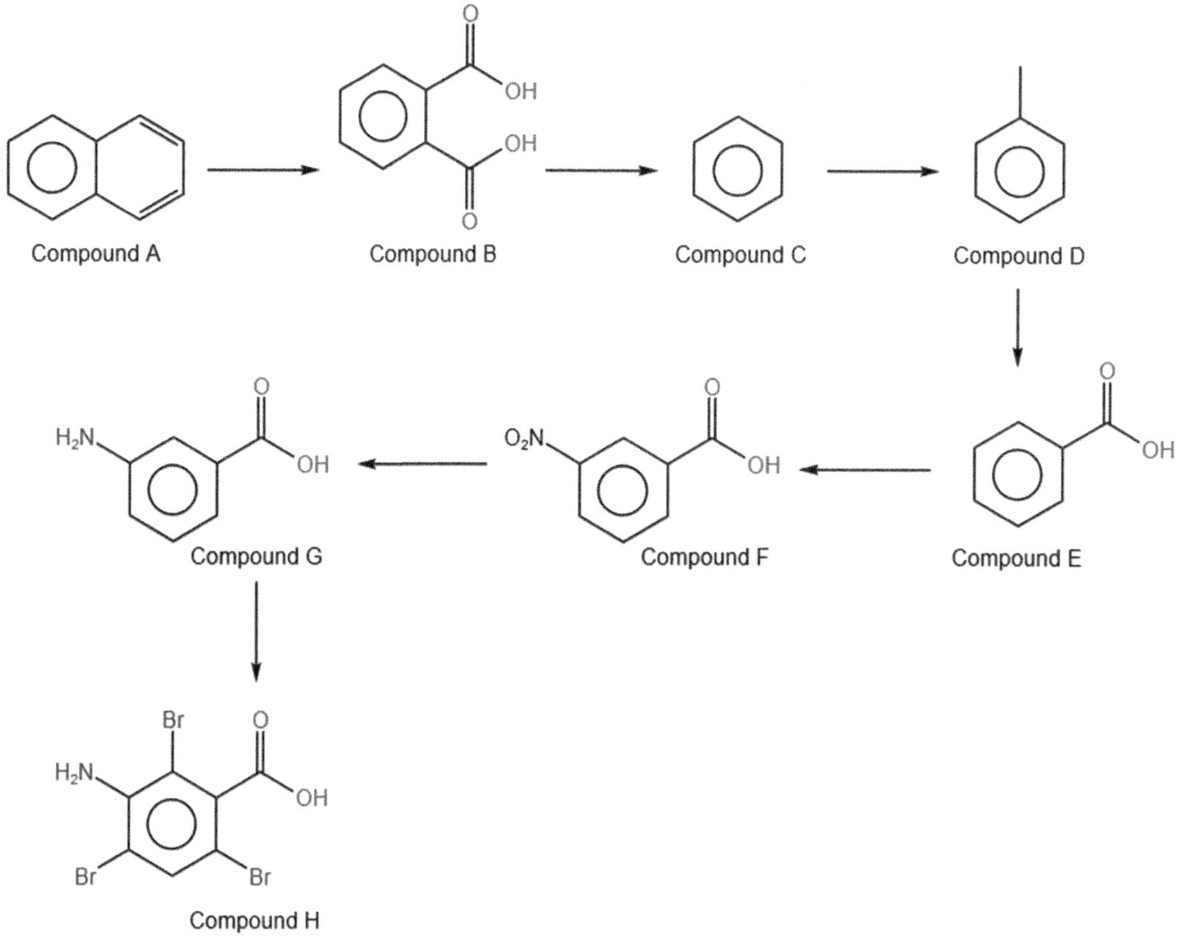
Compound A
OH
OH
Compound B
Compound C
Compound D
OH
Compound E
O2N
OH
Compound F
H2N
OH
Compound G
Br
H2N
OH
Br
Br
Compound H

Question 68:

Part (a) Compound **A** (C_8H_8) reacts with acidified potassium manganate(VII), leading to the decolourisation of the solution and forming **B** ($C_7H_6O_2$). Additionally, **A** reacts swiftly with liquid bromine, causing the solution to lose its color and resulting in the formation of **C** ($C_8H_8Br_2$).

When **C** is treated with aqueous sodium hydroxide, it undergoes a reaction to produce **D** ($C_8H_{10}O_2$). Upon reflux with acidified sodium dichromate(VI), **D** is transformed into **E** ($C_8H_6O_3$). Compound **E**, however, does not react with ammoniacal silver nitrate or Fehling's solution. It does, however, react with 2,4-dinitrophenylhydrazine, forming an orange-colored compound identified as **F** ($C_{14}H_{10}N_4O_6$).

Explain the observations at each step, detailing the transformations and identifying the functional groups or reactions involved in compounds **A** through **F**.

Part (b) Can potassium manganate(VII) solution be used to convert **D** into **E** instead of using acidified sodium dichromate(VI)? Provide a detailed explanation for your reasoning, considering the reactivity and selectivity of potassium manganate(VII) in oxidation reactions.

My thought process:

Part (a):
Compound **A** has a degree of unsaturation (DU) of 5, which immediately suggests the presence of a benzene ring since a benzene ring alone contributes 4 DU. Given that **A** contains only 8 carbon atoms, the extra unsaturation directs us toward a simple structure such as phenylethene. This example clearly shows how valuable the DU concept is in deducing structures. When phenylethene (compound **A**) undergoes side-chain oxidation, it forms benzoic acid (compound **B**). Additionally, the alkene group in phenylethene can participate in electrophilic addition to yield a dihalogenoalkane (compound **C**). This intermediate then reacts with NaOH through nucleophilic substitution, forming a diol (compound **D**). Upon refluxing with $Na_2Cr_2O_7$, the diol is oxidized to produce compound **E**, which has a DU of 6 and, with its three oxygen atoms, is interpreted as containing a phenyl ring, a ketone, and a carboxyl group. The fact that **E** neither reacts with Tollen's reagent nor with Fehling's solution indicates the presence of a ketone group, a conclusion further supported by a positive 2,4-DNPH test. Finally, the structure of compound **F** is derived from the condensation reaction between the ketone group in **E** and 2,4-DNPH.

Part (b):
Using $KMnO_4$ to convert compound **D** into **E** would not work as intended. If $KMnO_4$ were employed under these conditions, instead of getting compound **E**, the reaction would yield benzoic acid, accompanied by effervescence.

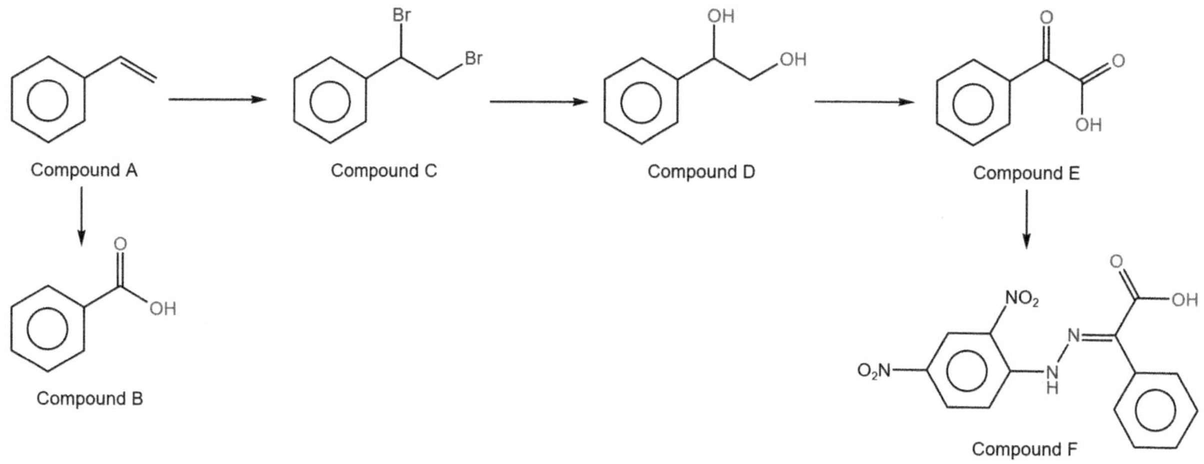
Compound A
Compound C
Compound D
Compound E
Compound B
Compound F

Question 69:

Part (a) Compound **A** ($C_{13}H_{12}$) reacts with acidified potassium manganate(VII) solution, causing decolourisation and resulting in the formation of **B** ($C_{13}H_{10}O$). Unlike some compounds, **A** does not decolourise bromine on its own but reacts under specific conditions with bromine to produce another product.

When **A** is exposed to chlorine in the presence of ultraviolet light, it forms **C** ($C_{13}H_{11}Cl$). With an excess of chlorine, **C** is further chlorinated to yield **D** ($C_{13}H_{10}Cl_2$). When **D** undergoes treatment with hot aqueous sodium hydroxide, **B** is generated. However, the structure of **B** differs from what might be anticipated based on the structure of **D**.

Compound **B** shows no reaction with ammoniacal silver nitrate or Fehling's solution. However, **B** reacts with 2,4-dinitrophenylhydrazine, producing an orange-colored derivative, **E** ($C_{19}H_{14}N_4O_4$). When **E** is treated with tin and excess concentrated hydrochloric acid, it is reduced to form a salt of **F** ($C_{19}H_{18}N_4$).

Identify compounds **A** through **F**, providing detailed explanations for the observed reactions and properties at each step. Support your answers with the relevant chemical principles.

Part (b) Draw and describe the structure of the product that would have been expected to form when **D** is treated with hot aqueous sodium hydroxide. Provide an explanation as to why this expected product does not form, incorporating the chemical reasoning behind this unexpected observation.

My thought process:

Part (a):

Compound **A** exhibits a degree of unsaturation of 8, and since degrees of unsaturation that are multiples of 4 often point to the presence of more than one phenyl ring, it is reasonable to suspect that two phenyl rings are present in its structure. When compound **A** is oxidized with $KMnO_4$, the product, compound **B**, is obtained with a degree of unsaturation increased to 9 and an extra oxygen atom incorporated into its structure. This additional oxygen atom strongly suggests that **B** is a ketone rather than an aldehyde, because under these conditions $KMnO_4$ typically does not yield aldehydes. The identification of the ketone functionality in **B** is further supported by its positive reaction with 2,4-dinitrophenylhydrazine (2,4-DNPH), a reagent that specifically detects carbonyl groups. Knowing the structure of **B**, which contains both the ketone group and two phenyl rings, lays the groundwork for deducing the structures of the subsequent compounds. Compound **E** is formed through a condensation reaction between the ketone group of **B** and 2,4-DNPH, leading to the formation of an imine. This intermediate, **E**, is then reduced using tin and concentrated hydrochloric acid, during which the nitrobenzene moiety introduced in the condensation reaction is reduced to a phenylamine group, culminating in the formation of compound **F**. Additionally, compound **A** can undergo free radical substitution to produce compound **C**, or, when exposed to an excess of chlorine, yield a di-chlorinated compound **D**. An interesting observation arises from the reaction of **D** under alkaline hydrolysis: instead of yielding the expected product, it forms ketone **B**. This transformation hints at the involvement of a geminal diol intermediate.

Part (b):

A geminal diol is characterized by having two hydroxyl (–OH) groups attached to the same carbon atom. Such compounds are generally less thermodynamically stable due to the unfavorable accumulation of electron density and steric crowding at the carbon center. As a result, geminal diols tend to undergo a spontaneous dehydration process, in which they lose a molecule of water to form a carbonyl compound. In the context of the reaction sequence described, the geminal diol formed during the alkaline hydrolysis of compound D is not stable and readily dehydrates to form ketone **B**. Thermodynamically, this dehydration is favored because it leads to an increase in the system's entropy; the production of water increases the number of particles in the system, which according to the Gibbs free energy equation ($\Delta G = \Delta H - T\Delta S$), results in a more negative ΔG if the entropy change is positive and significant. Under conditions where the enthalpy change (ΔH) and the temperature (T) remain relatively constant, the positive entropy change (ΔS) is the driving force that promotes the spontaneous formation of the ketone from the geminal diol.

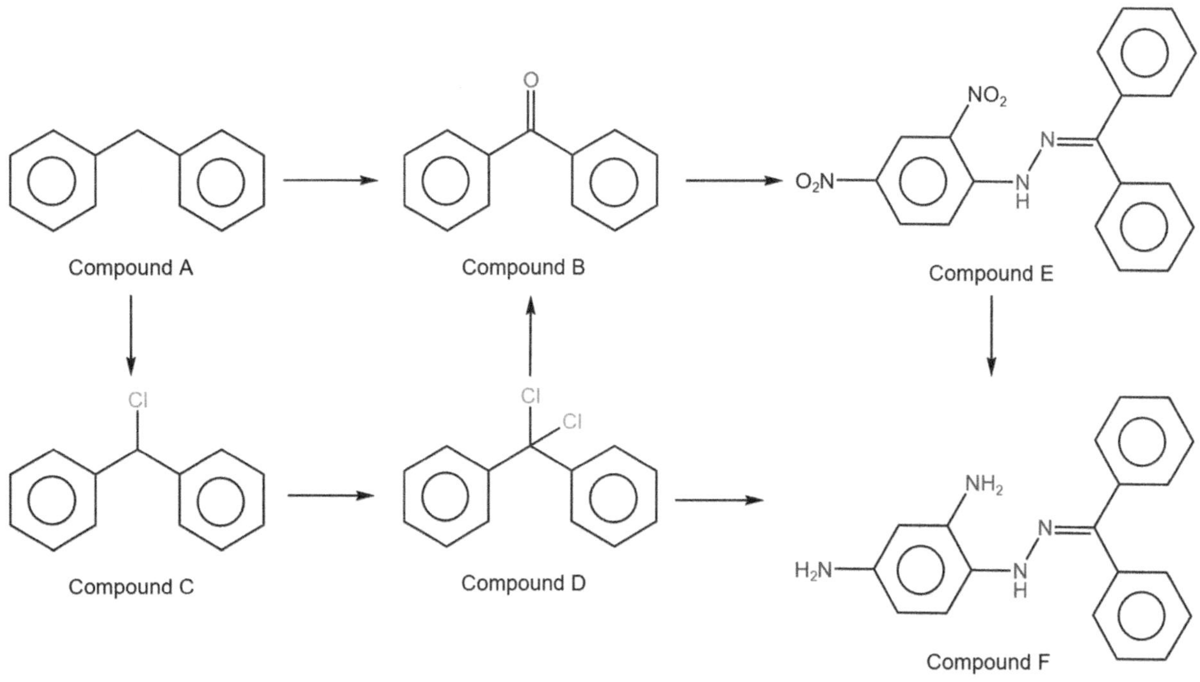

Question 70:

Compound **A** ($C_{19}H_{16}$) reacts with acidified potassium manganate(VII), leading to decolourisation of the solution and forming **B** ($C_{19}H_{16}O$). Despite this reactivity, **A** does not cause decolourisation of bromine. However, when exposed to bromine under specific conditions, **A** undergoes a reaction to produce a distinct product.

When **A** is treated with excess chlorine in the presence of ultraviolet light, it reacts to form **C** ($C_{19}H_{15}Cl$). Further treatment of **C** with hot alcoholic sodium cyanide leads to the formation of **D** ($C_{20}H_{15}N$). Acidification of **D** results in the generation of **E** ($C_{20}H_{16}O_2$).

Your task is to identify compounds **A** through **E**. For each compound, provide a detailed explanation of the observations, including the chemical transformations and principles that apply to these reactions. Ensure to support your identifications with relevant reasoning and chemical equations.

My thought process:

Compound **A** has the molecular formula $C_{19}H_{16}$, which yields a degree of unsaturation of 12. Because each phenyl ring contributes 4 units of unsaturation, the fact that the unsaturation is a multiple of 4 strongly suggests the presence of multiple aromatic rings. Combined with the carbon count, it is reasonable to suspect that **A** contains three phenyl rings. In this structure the three aromatic rings are bonded to a central carbon that also bears one hydrogen atom, a feature that becomes significant in subsequent reactions.

When compound **A** is oxidized using acidified potassium manganate(VII) ($KMnO_4$), the solution is decolorized and the product, compound **B**, is isolated with the formula $C_{19}H_{16}O$. The unchanged degree of unsaturation (12) together with the incorporation of one extra oxygen atom points to the introduction of a hydroxyl group. Thus, **B** is most likely triphenylmethanol, in which the benzylic hydrogen (the one attached to the central carbon in **A)** is oxidized to form an –OH group. The observation that **A** does not decolorize bromine confirms that it does not possess any carbon–carbon double bonds (alkene functionality); rather, the unsaturation results solely from the aromatic rings.

Additionally, when **A** undergoes free radical substitution with excess chlorine (in the presence of ultraviolet light), the reaction produces a mono-chlorinated compound, **C**, with the formula $C_{19}H_{15}Cl$. This selective substitution suggests that the only position available for such a radical reaction is the benzylic hydrogen attached to the central carbon—this carbon is already bonded to three phenyl rings and bears just one hydrogen, making it particularly susceptible to free radical halogenation. Thus, compound **C** is best identified as triphenylmethyl chloride.

In a further reaction, compound **C** is treated with hot alcoholic sodium cyanide (NaCN), resulting in the replacement of the chlorine atom with a cyanide group to yield compound **D**, which has the formula $C_{20}H_{15}N$. This nucleophilic substitution reaction is typical for benzylic halides, and the addition of the –CN group confirms that **D** is 2,2,2-triphenylethanenitrile, where the cyanide is attached to the benzylic carbon now bonded to three phenyl rings.

Finally, when compound **D** is acidified, it undergoes hydrolysis of the nitrile functionality. Under these conditions, the –CN group is converted to a carboxylic acid group, yielding compound **E** with the formula $C_{20}H_{16}O_2$. This transformation is consistent with the typical acid-catalyzed hydrolysis of nitriles to carboxylic acids, indicating that **E** is 2,2,2-triphenylethanoic acid.

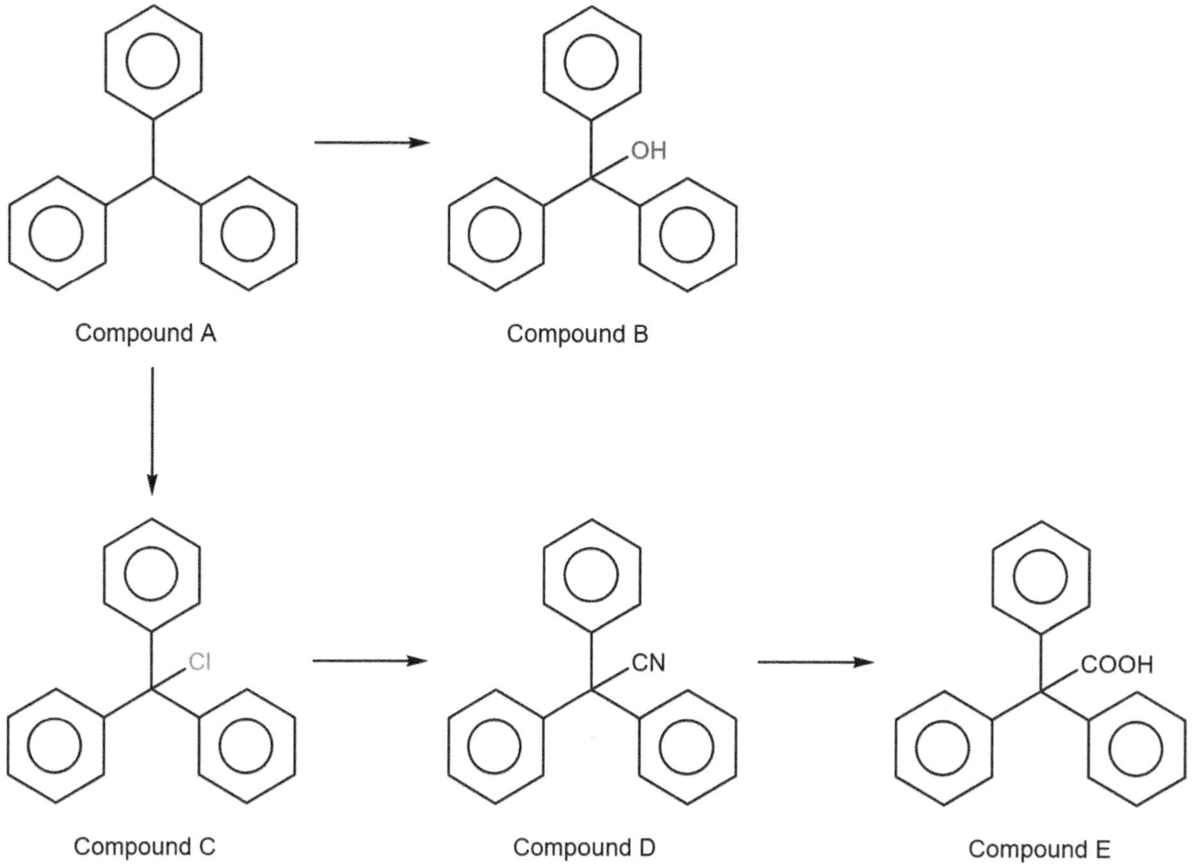

Question 71:

Compound **A** ($C_{16}H_{23}O_2N$) undergoes hydrogenation in the presence of a nickel catalyst, forming **B** ($C_{16}H_{25}O_2N$). Compound **B** reacts rapidly with bromine in tetrachloromethane, causing the solution to lose its color and producing **C** ($C_{16}H_{24}O_2NBr$). Notably, **C** exists as a pair of optical isomers.

When **A** is oxidised using acidified potassium manganate(VII), it produces three distinct organic products: **E** ($C_8H_7O_4N$), **F** ($C_4H_{10}O$), and **G** (C_3H_6O). Upon reaction with ethanoyl chloride, **E** forms **I** ($C_{10}H_9O_5N$). Effervescence is observed when sodium metal is added to **F**, indicating its chemical reactivity. Compound **G**, when reacted with 2,4-dinitrophenylhydrazine, generates an orange-colored precipitate identified as **H**.

Additionally, when 0.01 mol of **E** reacts with an excess of sodium metal, it liberates 0.24 dm^3 of hydrogen gas at room temperature and pressure. Furthermore, treating **A** with two moles of bromine in liquid form results in the formation of **D** ($C_{16}H_{22}O_2NBr_3$).

Your task is to identify compounds **A** through **I**, providing detailed explanations for all observed reactions and properties. Be sure to support your conclusions with the relevant chemical principles and equations.

My thought process:

Compound **A** has a degree of unsaturation equal to six, which strongly suggests the presence of a benzene ring (accounting for four degrees of unsaturation). The remaining two degrees may arise from the presence of an alkene group, a carbonyl (-C=O) group, or a combination of both. When **A** is treated with hydrogen gas in the presence of a nickel catalyst—a condition known to reduce alkenes and simple carbonyls (such as aldehydes and ketones) —these groups would be expected to undergo reduction. However, carboxylic acids and their derivatives are not affected under these conditions; their reduction typically requires lithium aluminium hydride ($LiAlH_4$) in dry ether.

The product of this hydrogenation, compound **B**, is found to decolourise bromine water, a test for unsaturation or activated aromatic rings. While this reaction may suggest the presence of an alkene, this possibility is ruled out because any alkene in **A** would have been reduced under the hydrogenation conditions. Therefore, the ability of **B** to decolourise bromine likely arises from the presence of a phenol or a phenylamine group.

When compound **A** is oxidised with potassium manganate(VII) ($KMnO_4$), three distinct products are obtained. Two of these, compounds **F** and **G**, can be readily identified. Compound **F** has no degrees of unsaturation and contains one oxygen atom. As it resists further oxidation, it must be a tertiary alcohol. Given that **F** contains four carbon atoms, it is most likely 2-methylpropan-2-ol.

Compound **G** has one degree of unsaturation and also contains one oxygen atom. This corresponds to a ketone, and the simplest possibility is propanone.

The third oxidation product, compound **E**, retains six degrees of unsaturation, consistent with the presence of an aromatic ring. It also contains four oxygen atoms and releases 0.01 mol of hydrogen gas per 0.01 mol of **E** when treated with sodium metal, suggesting the presence of two acidic –OH groups. This observation points to two carboxylic acid groups. The total degree of unsaturation in **E** aligns with this: four from the aromatic ring and two from the two carboxyl groups. Additionally, **E** is known to react with ethanoyl chloride in a condensation reaction, confirming that it contains an amino group. At this point, whether the amine is aliphatic or aromatic remains uncertain.

A key clue lies in the bromination of compound **A**. One mole of **A** is found to decolourise two moles of liquid bromine, and the resulting compound **D** contains three bromine atoms. This implies that two bromine atoms added across a C=C double bond, while one bromine atom substituted onto the aromatic ring. This substitution pattern is characteristic of a phenylamine, as the $-NH_2$ group is an activating, ortho/para-directing group. The fact that only one ring-position is brominated implies that the other two ortho/para positions are already occupied—most likely by two carboxyl groups.

Putting all this evidence together, compound **E** is best described as a substituted phenylamine with carboxylic acid groups at the 2- and 4- positions relative to the amino group.

Finally, to deduce the structure of compound **A**, we consider that its oxidation produces both a tertiary alcohol (**F**) and a ketone (**G**). This means that **A** must have contained ester groups that, upon hydrolysis (or side chain oxidation), yield these products (**F**, for instance). In addition, **A** must have an alkene group, which accounts for the bromine addition and its eventual hydrogenation.

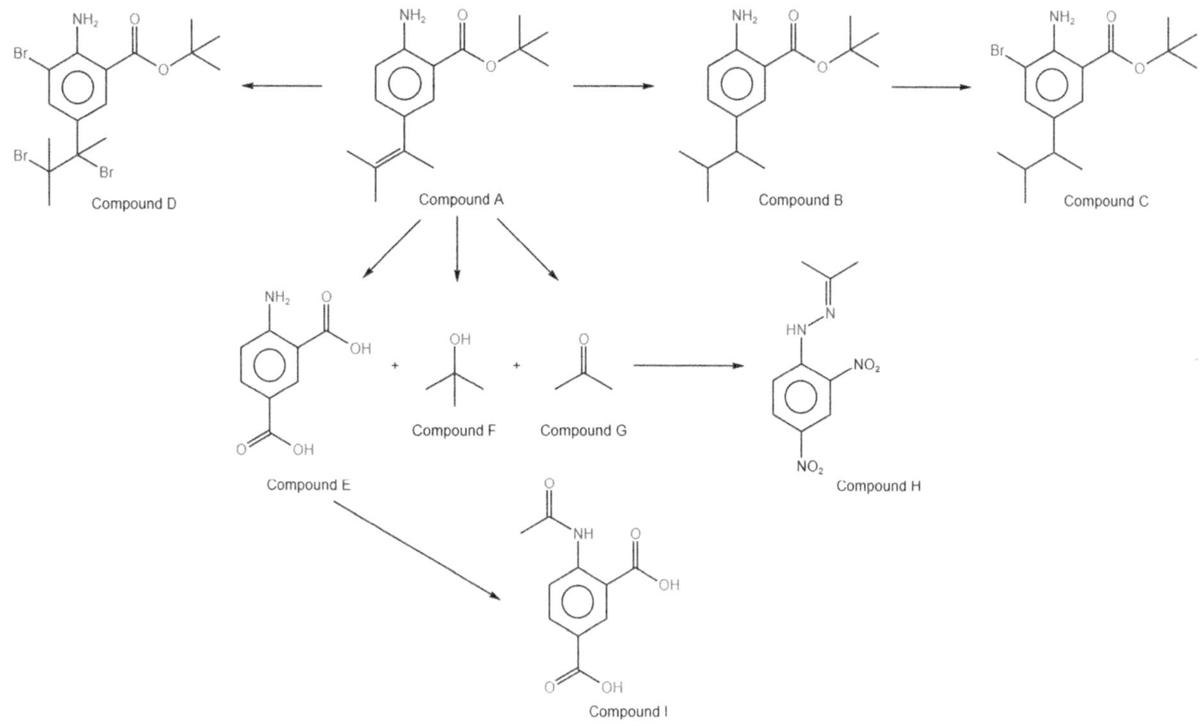
NH₂
O
O
Br
Compound A
Compound B
Compound C
Compound D
OH
Compound E
Compound F
Compound G
HN
N
NO₂
Compound H
NH
Compound I

Question 72:

Compound **A** (C_3H_6ONCl) is sparingly soluble in water but dissolves when treated with hydrochloric acid, forming **B** ($C_3H_5O_2Cl$). Adding sodium hydrogen carbonate to **B** causes effervescence, indicating the release of carbon dioxide. **B** dissolves in ethanolic potassium hydroxide, producing **C** ($C_3H_4O_2$). However, when **C** is treated with acidified potassium manganate(VII), effervescence is observed, but no organic product is formed.

When **B** is reacted with aqueous potassium hydroxide, followed by acidification, it yields **D** ($C_3H_6O_3$). Upon oxidation with acidified potassium manganate(VII), **D** is converted into **E** ($C_3H_4O_3$). Compound **E** reacts with 2,4-dinitrophenylhydrazine, forming an orange-yellow compound identified as **F** ($C_9H_8O_6N_4$). Further reaction of **F** with tin and excess concentrated hydrochloric acid, followed by alkaline hydrolysis, results in the formation of **G** ($C_9H_{12}O_2N_4$).

Your task is to determine the identities of **A** through **G**, providing comprehensive explanations for all observed reactions and transformations. Ensure to support your conclusions with detailed reasoning and appropriate chemical equations.

My thought process:

The conversion of **A** to **B** is accompanied by loss of nitrogen, which immediately suggests that **A** is an amide that, under acidic hydrolysis, gives **B**—a carboxylic acid. Both **A** and **B** display one degree of unsaturation, consistent with a single C=O group in each. With only three carbon atoms present, **A** (and therefore **B**) can exist as either the primary or the secondary chloro-isomer.

The carboxyl functionality of **B** is confirmed by vigorous effervescence with $NaHCO_3$ (to give $CO_2 + H_2O$), reinforcing the idea that **A** was both an amide and a halogenoalkane, and that **B** is a halogen-substituted carboxylic acid. When **B** undergoes elimination, it yields C, which contains an alkene plus a carboxyl group (DU = 2).

Treating **B** with aqueous KOH effects nucleophilic substitution of Cl for OH, and subsequent acidification protonates the carboxylate to give **D**—a carboxylic acid. At this stage it isn't yet clear whether the new –OH is primary or secondary. However, when **D** is subjected to strong $KMnO_4$ oxidation, the only possible product is a ketone (aldehydes would be further oxidised by $KMnO_4$), so **D** must have been a secondary alcohol. This, in turn, tells us that **A** and **B** were the secondary alkyl chloride isomers.

Finally, **E** (the ketone from oxidising D) reacts with 2,4-DNPH in a condensation to form **F**, and reducing F's nitro group (introduced by the 2,4-DNPH reagent) with Sn/HCl regenerates the amino functionality.

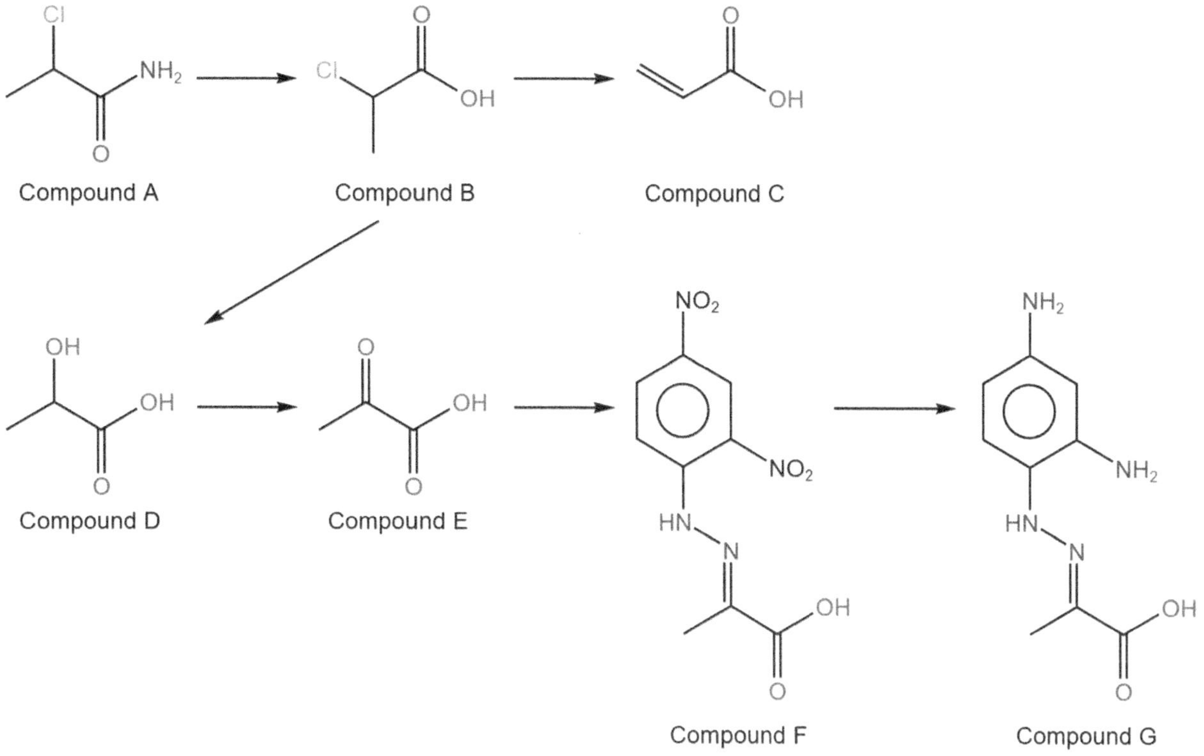
Cl
NH2
O
Compound A
Cl
O
OH
Compound B
O
OH
Compound C
OH
OH
O
Compound D
O
OH
O
Compound E
NO2
NO2
HN
N
OH
O
Compound F
NH2
NH2
HN
N
OH
O
Compound G

Question 73:

Compound **A** (C_4H_8ONCl) is only sparingly soluble in water but dissolves readily in hot aqueous sodium hydroxide, to yield two products: **B** ($C_2H_4O_2$) and **C** (C_2H_7NO). When sodium hydrogen carbonate is added to **B**, effervescence occurs, indicating the release of carbon dioxide gas.

Compound **C**, when treated with acidified potassium manganate(VII), undergoes oxidation, resulting in the formation of **D** ($C_2H_5NO_2$). Notably, **D** exists as a dipolar species, reflecting its zwitterionic structure.

Your task is to identify compounds **A** through **D**, providing detailed explanations for all observed chemical reactions, physical properties, and transformations. Be sure to support your identifications with sound chemical reasoning and balanced equations where appropriate.

My thought process:

When compound **A** is treated with hot aqueous sodium hydroxide, it yields two products. This strongly suggests that **A** undergoes alkaline hydrolysis, most likely of an amide. If A were an ester, both hydrolysis products would each contain an oxygen atom, making the total oxygen count two, which isn't consistent here.

One of the hydrolysis products, compound **B**, has one degree of unsaturation, contains two oxygen atoms, and produces effervescence with sodium hydrogen carbonate. This behaviour confirms the presence of a carboxylic acid group. Since **B** has only two carbon atoms, it must be ethanoic acid. This supports the conclusion that **A** is an amide, as alkaline hydrolysis of an amide produces a carboxylic acid (B) and an amine (C).

Compound **C** has no degrees of unsaturation and contains a hydroxyl group, suggesting it is a saturated amino alcohol. When **C** is oxidised with potassium manganate(VII), it forms compound **D**, which has one degree of unsaturation and two oxygen atoms. These features indicate that **D** is a carboxylic acid, confirming that **C** was a primary alcohol (as $KMnO_4$ oxidises primary alcohols to acids).

From this sequence, the structures of both **C** and **D** can be determined, leading naturally to the identification of compound **A**.

However, it is important to note that compound **C** cannot be CH_3–NH–CH_2OH. While this structure does contain both a primary alcohol and an amine, and would indeed be oxidised to form CH_3–NH–COOH (matching the molecular formula $C_2H_5NO_2$), under the reaction conditions—strong oxidation with $KMnO_4$ and heating with dilute sulfuric acid—this compound would undergo acidic hydrolysis of the amide linkage. As a result, CH_3–NH–COOH would break down further to give methylamine (CH_3NH_2), carbon dioxide, and water.

This instability under the reaction conditions rules out $CH_3–NH–CH_2OH$ as the correct structure for **C**, supporting a different amino alcohol that aligns with the observed stability and reaction outcomes.

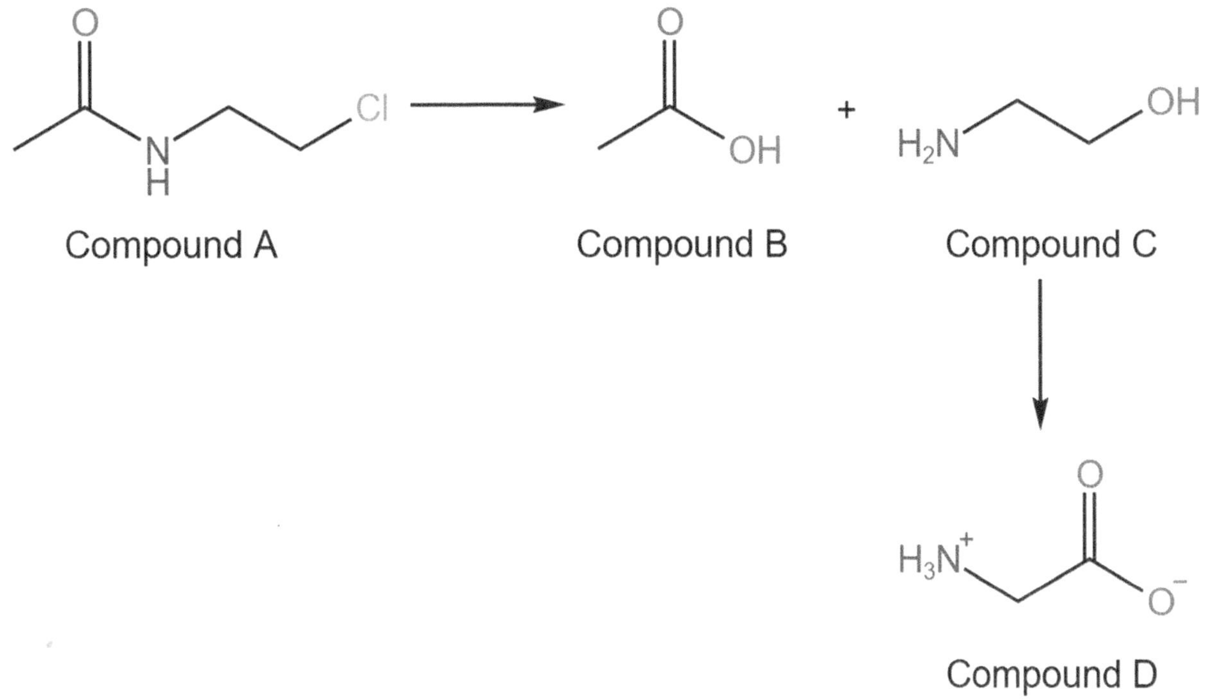

Question 74:

Compound **A** (C_6H_{12}) shows a single peak in its ^{1}H NMR spectrum, which indicates that all hydrogens in the molecule are chemically equivalent due to its symmetrical structure. **A** decolorizes bromine water. In a separate procedure, refluxing **A** with acidified potassium permanganate oxidizes it to give compound **B**.

A reacts with bromine water and acidified potassium manganate(VII), leading to decolourisation and the formation of **B** respectively.

When **B** is treated with 2,4-dinitrophenylhydrazine, it produces an orange-colored derivative identified as **C** ($C_9H_{10}N_4O_4$). Reacting **C** with tin and excess concentrated hydrochloric acid, followed by subsequent alkaline hydrolysis, converts it into **D** ($C_9H_{14}N_4$).

Additionally, when **B** is reacted with bromoethane in the presence of magnesium in dry ether, it results in the formation of **E** ($C_5H_{12}O$). Compound **E** does not react with acidified potassium manganate(VII), as no decolourisation occurs.

Additional Insight:

In ^{1}H NMR spectroscopy, the presence of a single peak is a unique feature observed in molecules where all hydrogen atoms are equivalent due to molecular symmetry. Examples of such molecules include methane (CH_4) and tetramethylsilane ($Si(CH_3)_4$), where identical environments produce a single resonance signal.

Your task is to identify compounds **A** through **E** and explain all observations related to their reactions and properties, incorporating chemical principles and balanced equations to support your reasoning.

My thought process:

Compound **A** has a degree of unsaturation equal to one and is highly symmetrical, as indicated by the presence of only a single peak in its ^{1}H NMR spectrum. Although spectroscopy is not part of the H2 Chemistry syllabus, students shouldn't be intimidated—questions that involve such topics typically include guiding hints to help bridge any gaps in knowledge.

The fact that **A** decolourises bromine suggests the presence of a C=C double bond, and since its degree of unsaturation is one, this confirms that **A** is an alkene with only one double bond. Moreover, oxidative cleavage of **A** using potassium manganate(VII) produces only one product, **B**, which reinforces the idea that **A** is highly symmetrical. These clues point clearly to the structure of **A** being 2,3-dimethylbut-2-ene, and **B**, the product of oxidative cleavage, is propanone.

Compound **B** (propanone) reacts with 2,4-dinitrophenylhydrazine (2,4-DNPH) to form compound **C** through a condensation reaction (an addition–elimination mechanism). The resulting product contains nitro groups, which came from the 2,4-DNPH reagent. Upon reduction with tin and concentrated hydrochloric acid, these nitro groups are converted into amino groups, forming phenylamine.

Separately, when bromoethane is reacted with magnesium in dry ether, it forms a Grignard reagent. Normally, the carbon atom in an alkyl halide is electrophilic due to the electron-withdrawing nature of the halogen atom. However, the presence of magnesium in dry ether reverses this polarity, generating a nucleophilic carbon—in this case, the ethyl group behaves like a carbanion ($CH_3CH_2^-$).

This Grignard reagent can now perform a nucleophilic addition to the carbonyl carbon of compound **B** (propanone), forming a tertiary alcohol, compound **E**. Because **E** is a tertiary alcohol, it cannot be further oxidised by potassium manganate(VII), which explains its resistance to oxidation.

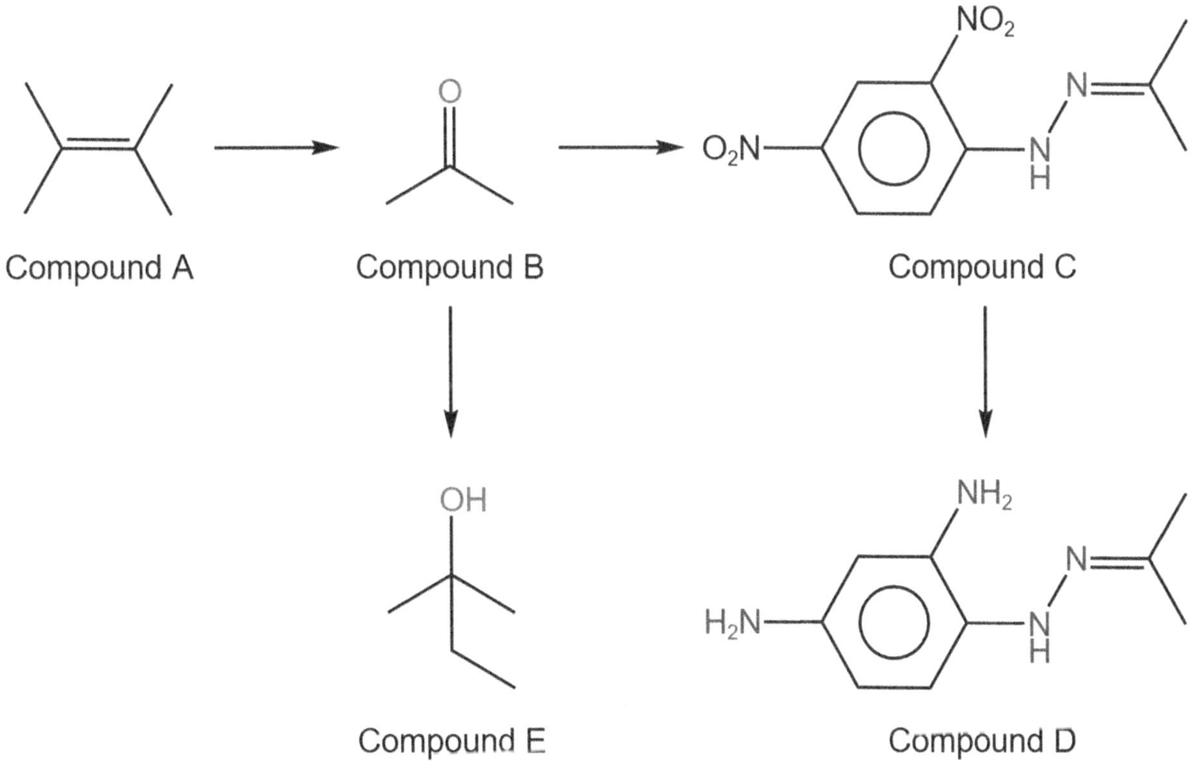

Question 75:

Compound **A** (C_6H_{12}) exhibits a single peak in its ^1H NMR spectrum, which indicates that all hydrogen atoms are in an identical chemical environment. This suggests a high degree of molecular symmetry. **A** does not decolourise bromine water. However, when exposed to chlorine in the presence of ultraviolet light, it forms **B** ($C_6H_{11}Cl$).

Compound **B** reacts with aqueous sodium hydroxide, leading to the formation of **C** ($C_6H_{12}O$). When **C** is heated with acidified potassium manganate(VII), it undergoes oxidation to produce **D**, a compound that gives an orange precipitate with 2,4-dinitrophenylhydrazine.

Reaction of **D** with bromoethane in the presence of magnesium in dry ether leads to the formation of **E** ($C_8H_{16}O$), which does not react with acidified potassium manganate(VII). Passing hydrogen cyanide gas through **D** results in the formation of **F** ($C_7H_{11}ON$), which, when acidified, converts to **G** ($C_7H_{12}O_3$).

When **B** is heated with alcoholic sodium hydroxide, **H** (C_6H_{10}) is generated. Oxidation of **H** with acidified potassium manganate(VII) yields **I** ($C_6H_{10}O_4$). Treating **I** with phosphorus pentachloride results in the release of steamy fumes and the formation of **J** ($C_6H_8Cl_2O_2$). Finally, when **J** is treated with ammonia, it reacts to form **K** ($C_6H_{12}O_2N_2$).

Your task is to identify compounds **A** through **K**, providing detailed explanations for each observed reaction and transformation. Be sure to support your answers with logical reasoning and balanced chemical equations.

My thought process:

Compound **A** has a degree of unsaturation equal to one and is highly symmetrical, as suggested by the presence of only a single peak in its ^1H NMR spectrum. Since **A** does not decolourise bromine water, it is not an alkene. However, it undergoes free radical substitution (FRS) with chlorine gas to form compound **B**, a halogenoalkane. This reactivity points to **A** being a cycloalkane, specifically cyclohexane—a symmetrical compound with one ring (one degree of unsaturation).

Once **A** is confirmed as cyclohexane, the subsequent transformations from **B** to **K** can be worked out systematically:

- **B** (chlorocyclohexane) is formed via FRS.
- **B** undergoes nucleophilic substitution with aqueous NaOH to give **C** (cyclohexanol), an alcohol.

- **C** is oxidised by $KMnO_4$ to form **D** (cyclohexanone). The fact that **D** gives a positive 2,4-DNPH test confirms it is a ketone. This also supports the conclusion that **C** is a secondary alcohol.

Next, **D** (cyclohexanone) reacts with a Grignard reagent (ethylmagnesium bromide, effectively supplying an ethyl carbanion, $CH_3CH_2^-$) in a nucleophilic addition reaction to form **E**, a tertiary alcohol. Since tertiary alcohols resist oxidation, this explains why **E** does not react with potassium manganate(VII).

In another pathway, **D** undergoes nucleophilic addition with hydrogen cyanide (HCN) to produce **F**, a cyanohydrin. Upon acidic hydrolysis of the nitrile group in **F**, the resulting product is **G**, an α-hydroxycarboxylic acid.

Returning to **B** (chlorocyclohexane), it can also undergo elimination with alcoholic KOH, yielding **H** (cyclohexene). Cyclohexene is then subjected to oxidative cleavage with hot $KMnO_4$, producing **I** (hexanedioic acid), a six-carbon dicarboxylic acid.

Finally, **I** reacts with phosphorus pentachloride (PCl_5) via nucleophilic acyl substitution, forming **J** (hexanedioyl dichloride). This acyl chloride then undergoes condensation with ammonia, resulting in **K**, a diamide.

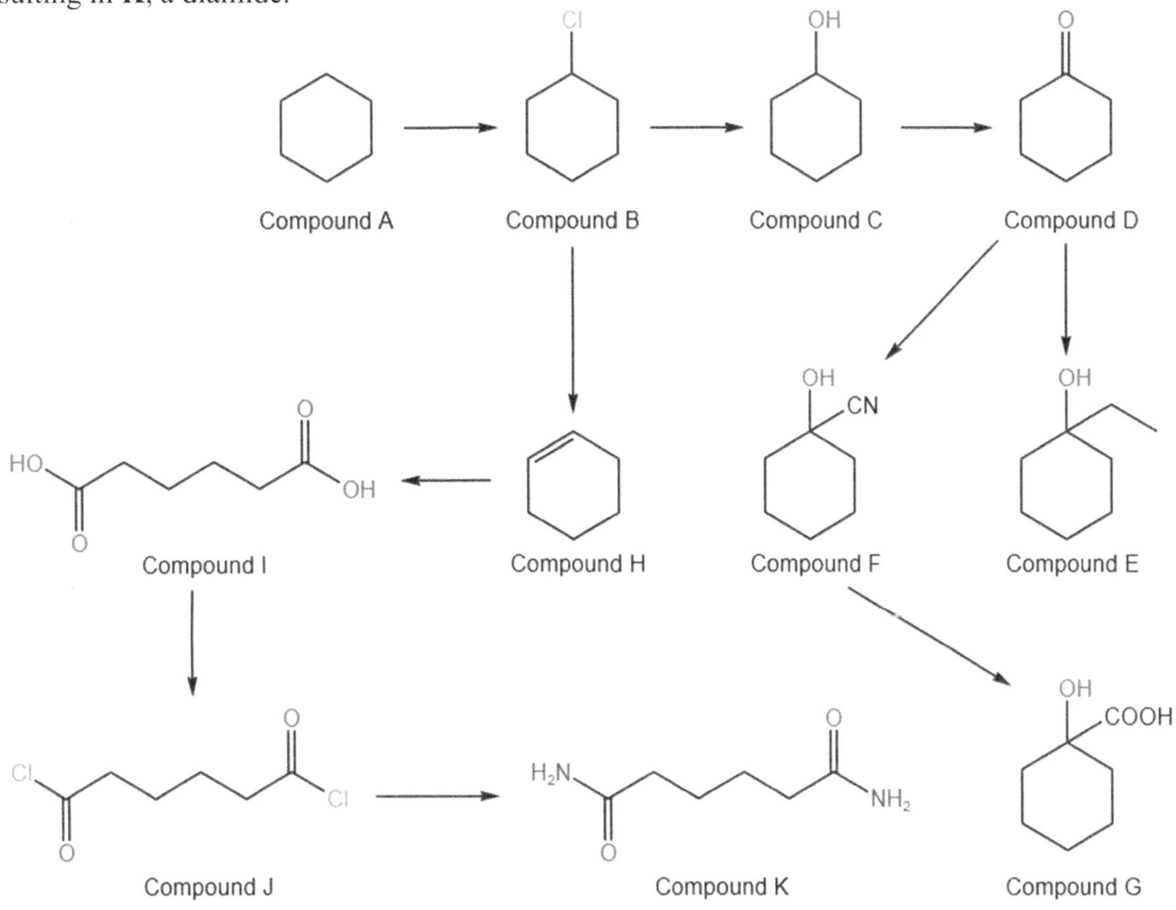

Question 76:

Compound **A** ($C_3H_6Cl_2$) exhibits optical activity, implying the presence of a chiral center in its structure. When treated with aqueous sodium hydroxide, **A** undergoes a reaction that results in the formation of **B** ($C_3H_8O_2$). Heating **B** with acidified potassium manganate(VII) causes an oxidation reaction, giving rise to **C** ($C_3H_4O_3$).

Using sodium borohydride in pyridine to reduce **C** produces **D** ($C_3H_6O_3$). When **D** is subjected to concentrated sulfuric acid at 170°C, dehydration occurs, leading to the formation of **E** ($C_3H_4O_2$). The reduction of **E** with lithium aluminium hydride in dry ether transforms it into **F** (C_3H_6O).

Further oxidation of **F** with acidified potassium dichromate(VI) with immediate distillation results in **G** (C_3H_4O). Compound **G**, when exposed to hydrogen cyanide gas in the presence of trace sodium hydroxide, undergoes a reaction to form **H** (C_4H_5ON). Further treatment of **H** with anhydrous lithium aluminium hydride in dry ether yields **I** ($C_4H_{11}ON$).

Your task is to identify compounds **A** through **I**, giving detailed explanations of the processes and transformations involved. Include balanced chemical equations and scientific reasoning to support your answers.

My thought process:

This question can be approached efficiently if students are comfortable using skeletal formulae to visualise structural isomers. Given that compound **A** has a degree of unsaturation of zero, it must be a saturated dihalogenoalkane. The molecular formula $C_3H_6Cl_2$ corresponds to four possible isomers: 1,1-dichloropropane, 1,2-dichloropropane, 1,3-dichloropropane, and 2,2-dichloropropane. Out of these, only 1,2-dichloropropane contains a chiral centre, which explains why **A** is optically active. Thus, compound **A** is 1,2-dichloropropane.

When **A** is treated with aqueous sodium hydroxide, it undergoes nucleophilic substitution, with both chlorine atoms replaced by hydroxyl groups to give 1,2-dihydroxypropane, compound **B**. This diol contains both a primary and a secondary alcohol group. Upon oxidation with potassium manganate(VII), **B** is converted into compound **C**, which features a carboxylic acid (from the primary alcohol) and a ketone (from the secondary alcohol).

When sodium borohydride ($NaBH_4$) is used as the reducing agent, only the ketone group is reduced, as $NaBH_4$ is not strong enough to reduce carboxylic acids. For that transformation, lithium aluminium hydride ($LiAlH_4$) in dry ether would be needed.

An interesting twist occurs during the reaction of compound **G** with hydrogen cyanide (HCN). Compound **G** contains an α,β-unsaturated carbonyl group, and this setup introduces an important conjugation between the π electrons of the C=C double bond and the C=O group. Instead of the expected nucleophilic attack at the carbonyl carbon, the cyanide ion (CN^-) preferentially attacks the β-carbon of the alkene. This is due to the resonance delocalisation of electron density, which makes the tail-end of the double bond more electrophilic than the carbonyl carbon itself. This type of conjugate (1,4-) addition is a classic feature of α,β-unsaturated systems and has appeared in previous A-level exam questions.

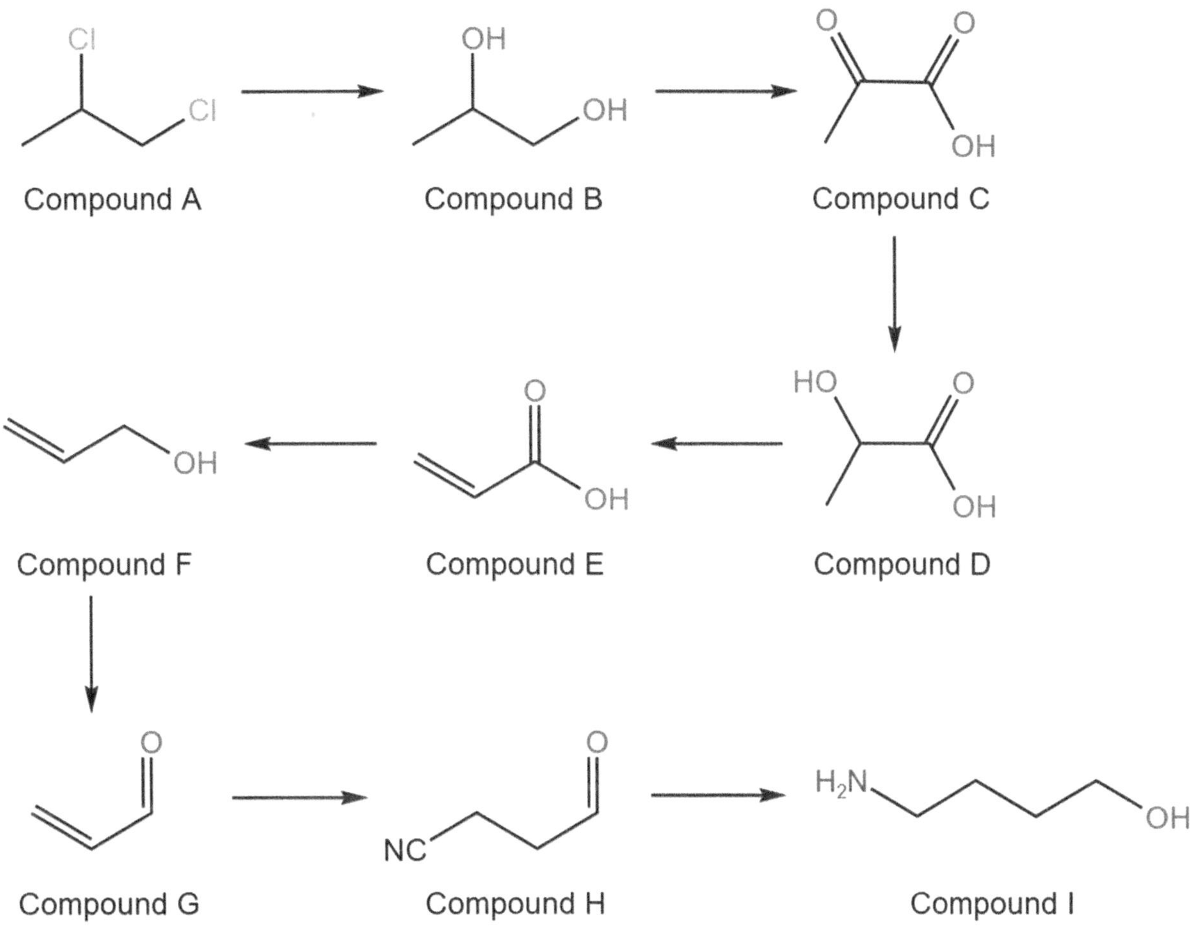

Question 77:

Compound **A** ($C_3H_6Cl_2$) lacks enantiomerism, indicating the molecule is not chiral and does not have non-superimposable mirror images. When **A** reacts with aqueous sodium hydroxide, the expected product **B** ($C_3H_8O_2$) is surprisingly not formed. Instead, the reaction produces **C** (C_3H_6O), which undergoes positive tests with 2,4-dinitrophenylhydrazine, ammoniacal silver nitrate, and Fehling's solution.

In the presence of an acid catalyst, two molecules of **C** combine, releasing one molecule of water, and forming **D** ($C_6H_{10}O$). Oxidation of **D** with acidified sodium dichromate(VI) yields **E** ($C_6H_{10}O_2$). When **D** is instead treated with hot acidified potassium manganate(VII), it is oxidized to form **F** ($C_3H_4O_3$), along with a secondary product, propanoic acid.

Further examination of **F** reveals its reaction with warm alkaline aqueous iodine. Upon acidification, the mixture produces the organic compound **G**, accompanied by the formation of CHI_3 (iodoform). Additionally, **F** reacts with 2,4-dinitrophenylhydrazine, producing an orange precipitate identified as **H** ($C_9H_8O_6N_4$).

Your task is to identify the structures of **A** through **H**. Provide detailed explanations for all reactions and observations, incorporating chemical reasoning and balanced equations to substantiate your conclusions.

My thought process:

This question becomes much more approachable if students are confident in drawing and interpreting skeletal formulae, especially when identifying structural isomers. Given that compound **A** has no degrees of unsaturation, it must be a fully saturated dihalogenoalkane. The molecular formula $C_3H_6Cl_2$ corresponds to four isomers: 1,1-dichloropropane, 1,2-dichloropropane, 1,3-dichloropropane, and 2,2-dichloropropane. Since **A** is not optically active, we can eliminate 1,2-dichloropropane (the only chiral isomer), leaving the remaining three as possibilities.

When **A** is treated with aqueous sodium hydroxide, we expect a nucleophilic substitution, replacing both chlorine atoms with hydroxyl groups to form a diol (compound **B**). However, the actual product observed is compound **C**, an aldehyde, as confirmed by positive reactions with 2,4-DNPH, Tollens', and Fehling's solutions. This suggests that compound **B** must be a geminal diol—a compound where both –OH groups are attached to the same carbon. Geminal diols are generally unstable and spontaneously dehydrate to form a carbonyl compound, which in this case is an aldehyde.

This key observation narrows down the identity of **A** further: only 1,1-dichloropropane and 2,2-dichloropropane can form geminal diols upon hydrolysis. However, dehydration of

2,2-propanediol would produce a ketone, whereas dehydration of 1,1-propanediol gives an aldehyde, aligning with the identity of **C**. Therefore, compound **A** must be 1,1-dichloropropane.

Once **A** is confirmed, the transformations follow logically:

- **A** undergoes nucleophilic substitution to give geminal diol **B**.
- **B** spontaneously dehydrates to give **C** (propanal).

Next, when two molecules of **C** (an aldehyde) react under suitable conditions (either acid or base catalyzed), they undergo aldol condensation. Initially, this forms a β-hydroxy aldehyde (aldol), which then readily dehydrates to yield compound **D**, an α,β-unsaturated carbonyl compound. This dehydration is driven by increased entropy and a more negative Gibbs free energy, making the process spontaneous and thermodynamically favourable—similar to the dehydration observed earlier with the geminal diol.

This entire transformation is an example of a condensation reaction, more specifically known as the aldol condensation, which can also be described mechanistically as an addition–elimination process.

🔍 **Tip for students:** Aldol condensation is a high-frequency topic in A-level organic chemistry exams. Understanding the mechanism—especially the role of enolate formation, nucleophilic attack, and subsequent elimination—is essential for tackling questions like this with confidence.

Once compound **D** is identified as the α,β-unsaturated carbonyl product, deducing the remaining compounds **E** to **H** becomes much more straightforward and follows naturally from standard reaction pathways.

Base-catalyzed Aldol condensation Mechanism (FYI)

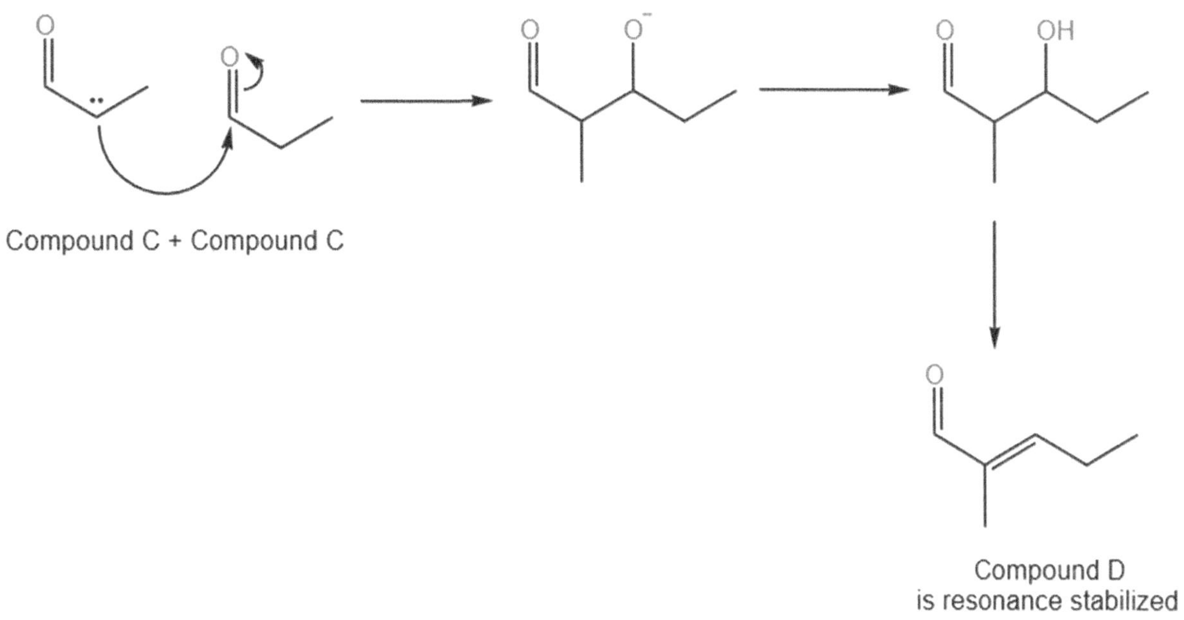

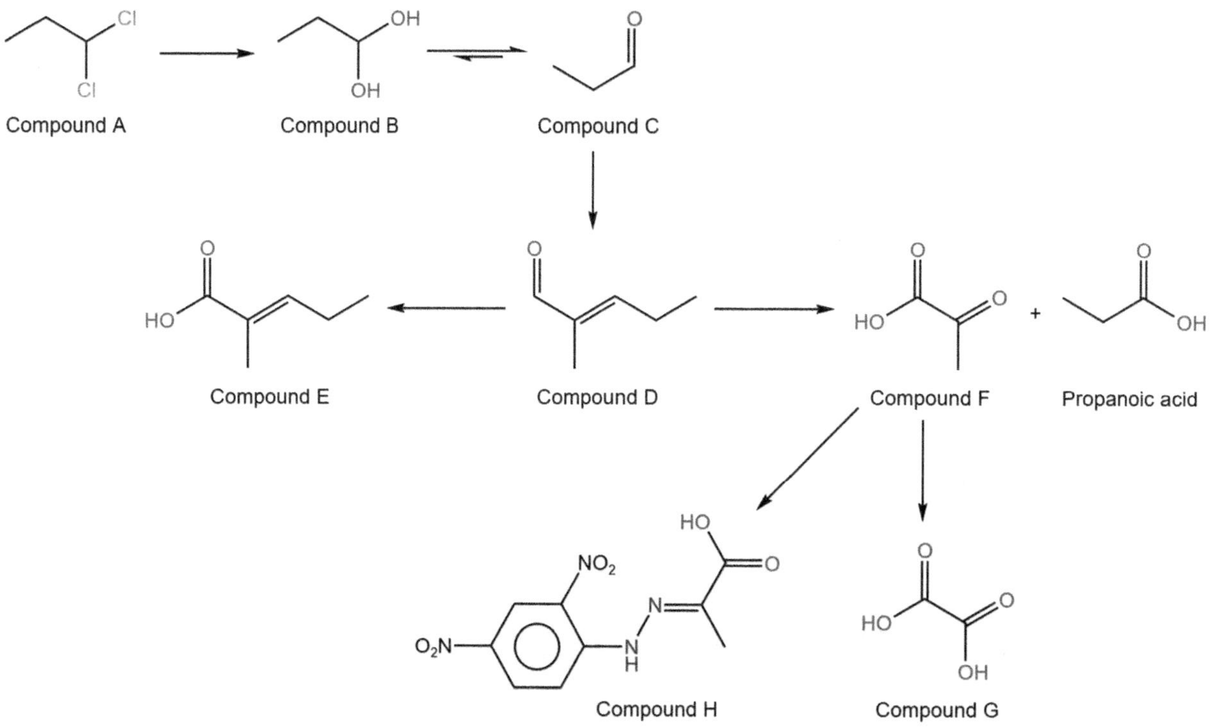
Compound A
Compound B
Compound C
Compound E
Compound D
Compound F
Propanoic acid
Compound H
Compound G

Question 78:

Compound **A** ($C_3H_6Cl_2$) is a symmetrical molecule that lacks chirality, meaning it is non-enantiomeric. When reacted with aqueous sodium hydroxide, A undergoes substitution and hydrolysis, forming **B** ($C_3H_8O_2$). Oxidizing **B** with acidified potassium manganate(VII) results in **C**, identified as $C_3H_4O_4$, a highly oxidized derivative.

If **B** is treated with acidified potassium dichromate(VI) and the reaction is distilled promptly, the product **D** ($C_3H_4O_2$) is obtained. Introducing cold hydrogen cyanide gas to **D** in the presence of trace sodium hydroxide induces nucleophilic addition, generating **E** ($C_5H_6N_2O_2$). Dissolving **E** in hot hydrochloric acid produces **F** ($C_5H_8O_6$), which undergoes reduction when treated with lithium aluminium hydride in dry ether, yielding **G** ($C_5H_{12}O_4$).

Remarkably, when sodium metal is added to **G**, the reaction releases twice the amount of hydrogen gas as there are moles of **G**, highlighting the molecular structure's reactive hydrogen atoms. Separately, reacting **A** with ethanolic potassium cyanide results in **H** ($C_5H_6N_2$), which, upon acidification, transforms into **I** ($C_5H_8O_4$).

Your challenge is to uncover the identities of compounds **A** through **I** while detailing the observations and chemical principles involved in each reaction. Ensure your answers are supported by logical deductions and balanced chemical equations.

My thought process:

This question becomes much more manageable when students are comfortable working with skeletal structures, especially for identifying structural isomers. Since compound **A** has no degrees of unsaturation, it must be a saturated di-halogenoalkane. The molecular formula $C_3H_6Cl_2$ allows for four possible isomers: 1,1-dichloropropane, 1,2-dichloropropane, 1,3-dichloropropane, and 2,2-dichloropropane. Given that **A** is not optically active, we can rule out 1,2-dichloropropane, as it is the only isomer that contains a chiral centre. This narrows down the options to the remaining three isomers.

Upon oxidation of compound **B** with acidified potassium manganate(VII), compound **C** is formed. **C** has a degree of unsaturation of two and contains four oxygen atoms—two more than **B**. This suggests that **C** possesses two carboxylic acid groups, identifying it as propanedioic acid. Knowing this, we can deduce that **B** must be 1,3-propanediol, which in turn confirms that **A** is 1,3-dichloropropane. Once compounds **A** and **B** are established, the identities of the remaining compounds from **D** to **I** can be determined with ease.

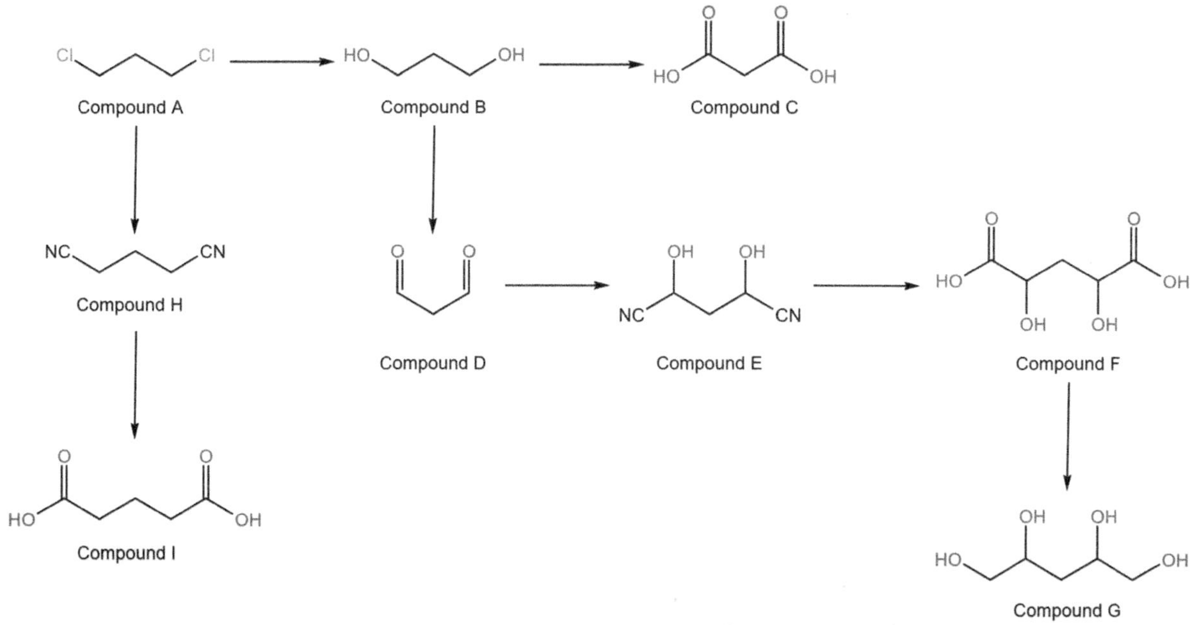
Cl
Cl
Compound A
HO
OH
Compound B
O
O
HO
OH
Compound C
NC
CN
Compound H
O
O
Compound D
OH
OH
NC
CN
Compound E
O
O
HO
OH
OH
OH
Compound F
O
O
HO
OH
Compound I
OH
OH
HO
OH
Compound G

Question 79:

Compound **A** ($C_3H_6Cl_2$) is non-chiral, confirming it does not exhibit optical isomerism. Treatment of **A** with aqueous sodium hydroxide does not yield the anticipated product **B** ($C_3H_8O_2$). Instead, the reaction produces **C** (C_3H_6O). Testing **C** with 2,4-dinitrophenylhydrazine gives a positive result, but it does not react with ammoniacal silver nitrate or Fehling's solution.

In the presence of an acid catalyst, two molecules of **C** condense with the loss of water to form **D** ($C_6H_{10}O$). When **D** is treated with hot acidified potassium dichromate(VI), there is no new compound produced, suggesting resistance to this oxidizing agent. However, when heated with acidified potassium manganate(VII), **D** undergoes oxidation to produce **E** ($C_3H_4O_3$), along with the regeneration of **C** as a by-product.

When **E** is treated with alkaline aqueous iodine, the reaction leads to the formation of **F**, an organic salt, accompanied by the production of CHI_3 (iodoform). Additionally, **E** reacts with phosphorus tribromide to form **G** ($C_3H_3BrO_2$). Further treatment of **G** with ammonia results in the formation of **H** ($C_3H_5NO_2$).

Your task is to identify the molecular structures of **A** through **H**, providing detailed explanations for each reaction. Ensure your reasoning is supported by clear chemical principles and balanced equations where applicable.

My thought process:

Students should take note of how Questions 76 to 79 share striking similarities—yet a subtle variation in the wording or given conditions can lead to entirely different outcomes. It's a strong reminder to pay close attention to these minor tweaks, as they often hold the key to solving the problem.

At this point, it should be clear that the molecular formula $C_3H_6Cl_2$ corresponds to four structural isomers: 1,1-dichloropropane, 1,2-dichloropropane, 1,3-dichloropropane, and 2,2-dichloropropane. Given that compound **A** is not optically active, we can immediately eliminate 1,2-dichloropropane, the only isomer that contains a chiral centre. This narrows down the possibilities to three non-chiral isomers.

Compound **B** is identified as a geminal diol, which means both hydroxyl groups are attached to the same carbon atom. Geminal diols are generally unstable and readily undergo dehydration to form carbonyl compounds. For **B** to be a geminal diol, **A** must have originally been either 1,1-dichloropropane or 2,2-dichloropropane, as only these two can form such structures upon hydrolysis.

Alkaline hydrolysis of 1,1-dichloropropane yields 1,1-propanediol, which upon dehydration gives an aldehyde (propanal). In contrast, hydrolysis of 2,2-dichloropropane gives 2,2-propanediol, which upon dehydration produces a ketone (propanone). Since compound **C** gives a positive test only with 2,4-DNPH and not with Tollens' or Fehling's reagent, it confirms that **C** is a ketone, not an aldehyde. This observation allows us to conclude that **C** is propanone, meaning **B** must be 2,2-propanediol, and hence **A** is 2,2-dichloropropane.

It's worth comparing this question (Q79) with Question 77, as they are structurally similar. However, the key difference lies in the nature of the carbonyl compound formed after the dehydration of the geminal diol—aldehyde versus ketone. This difference significantly alters the reaction pathway and the final products.

Next, the aldol condensation of two moles of carbonyl compound (in this case, two molecules of propanone) leads to the formation of compound **D**, an α,β-unsaturated ketone. Once students understand how aldol condensation proceeds mechanistically, identifying compound **D** becomes straightforward. From there, the structures of the subsequent compounds **E** to **H** follow logically through standard organic transformations.

Base-catalyzed Aldol condensation Mechanism (FYI)

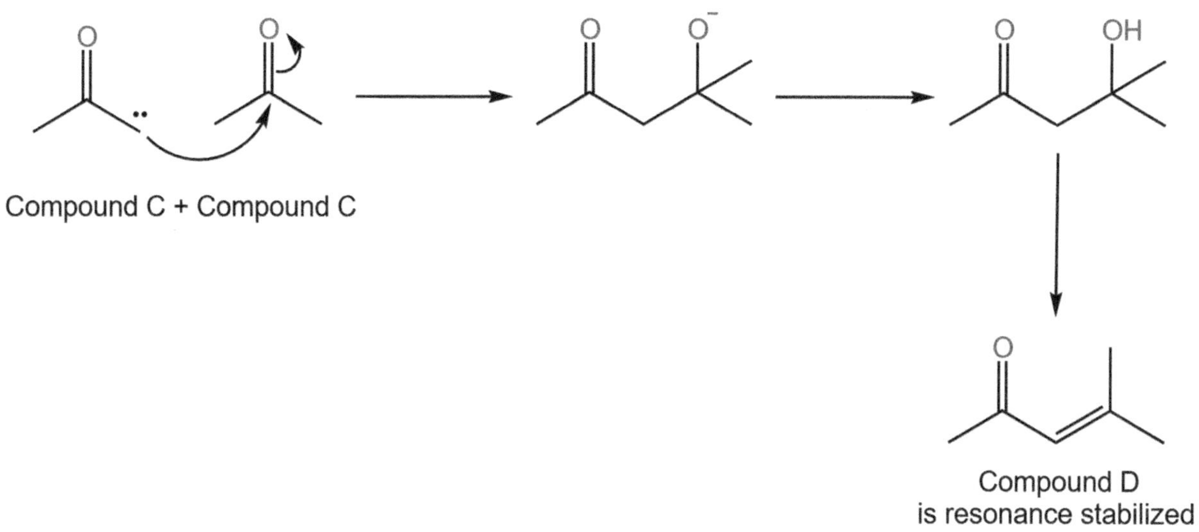

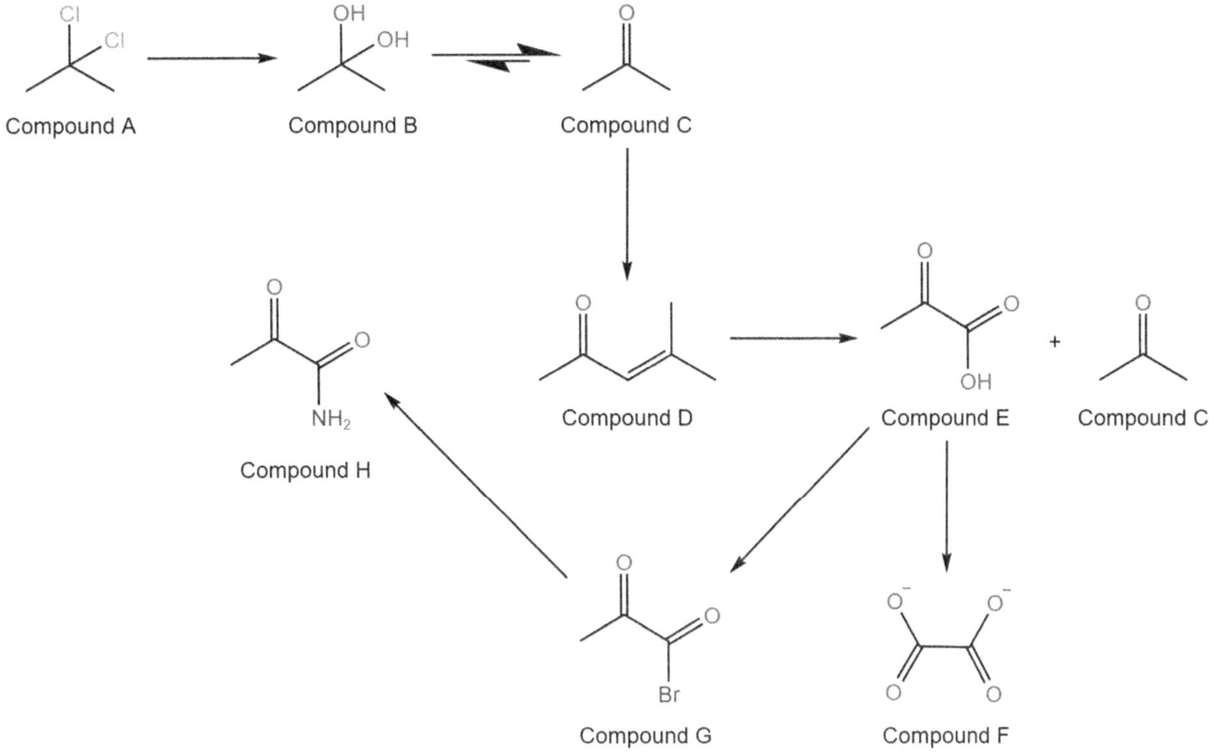
Cl
Cl
Compound A
OH
OH
Compound B
O
Compound C
O
Compound D
O
O
OH
Compound E
+
O
Compound C
O
O
NH2
Compound H
O
O
Br
Compound G
O⁻
O⁻
O
O
Compound F

Question 80:

Compound **A** (C_6H_{12}) produces a single peak in its ^{1}H NMR spectrum, which reflects the uniform chemical environment of all hydrogen atoms due to the molecular symmetry. **A** does not react with bromine water, indicating the absence of unsaturated bonds. Subjecting **A** to chlorine exposure under ultraviolet light produces **B** ($C_6H_{11}Cl$). When **B** reacts with aqueous sodium hydroxide under reflux conditions, it yields **C** ($C_6H_{12}O$), an alcohol derivative.

Heating **C** with acidified potassium manganate(VII) oxidizes it to form **D** ($C_6H_{10}O$). Treating **D** in the presence of sodium ethanoate and then acidifying the reaction mixture causes two molecules of **D** to combine, producing **E** ($C_{12}H_{18}O$), while eliminating one molecule of water. When heated with acidified potassium manganate(VII), **E** yields **F** ($C_6H_8O_2$) and regenerates **D** as a secondary product.

Hydrogen cyanide gas is bubbled through **F** in the presence of trace sodium hydroxide, resulting in the formation of **G** ($C_8H_{10}N_2O_2$). Acidification of **G** transforms it into **H**. Finally, **G** is subjected to hydrogen gas under elevated pressure in the presence of a nickel catalyst, reducing it to **I**.

Your task is to determine the identities of compounds **A** through **I**, providing detailed explanations for the observations recorded at each stage of the process. Include balanced chemical equations and logical reasoning to support your conclusions.

My thought process:

Compound **A** is a hydrocarbon with a degree of unsaturation equal to one. Since it shows only a single signal in the ^{1}H NMR spectrum and does not decolourise bromine water, this rules out the presence of a carbon–carbon double bond. These features strongly suggest that **A** is cyclohexane, where the unsaturation is accounted for by the ring itself, and the high symmetry of the molecule explains the single NMR peak.

When cyclohexane undergoes free radical substitution (FRS) with chlorine, the product is chlorocyclohexane (**B**). Treating **B** with aqueous sodium hydroxide results in nucleophilic substitution, forming cyclohexanol (**C**). Oxidation of **C** using acidified potassium manganate(VII) yields cyclohexanone (**D**).

This sequence closely parallels the one presented in Question 75, where students explore the formation and application of Grignard reagents. In that case, the Grignard reagent reacts with a ketone to form a tertiary alcohol. In contrast, the focus in this question shifts to base-catalysed enolate chemistry. Rather than introducing a Grignard reagent, a strong base such as sodium ethanoate is used to remove an acidic α-hydrogen from cyclohexanone. This generates an enolate ion, which then attacks another molecule of cyclohexanone to form an aldol product. Upon dehydration, this intermediate gives rise to an α,β-unsaturated ketone (compound **E**).

This type of reaction should be familiar by now, as it is the classic aldol condensation, also encountered in Questions 77 and 79. Once the structure of compound **E** is established, determining the identities of the subsequent compounds **F** through **I** becomes straightforward based on standard organic transformations.

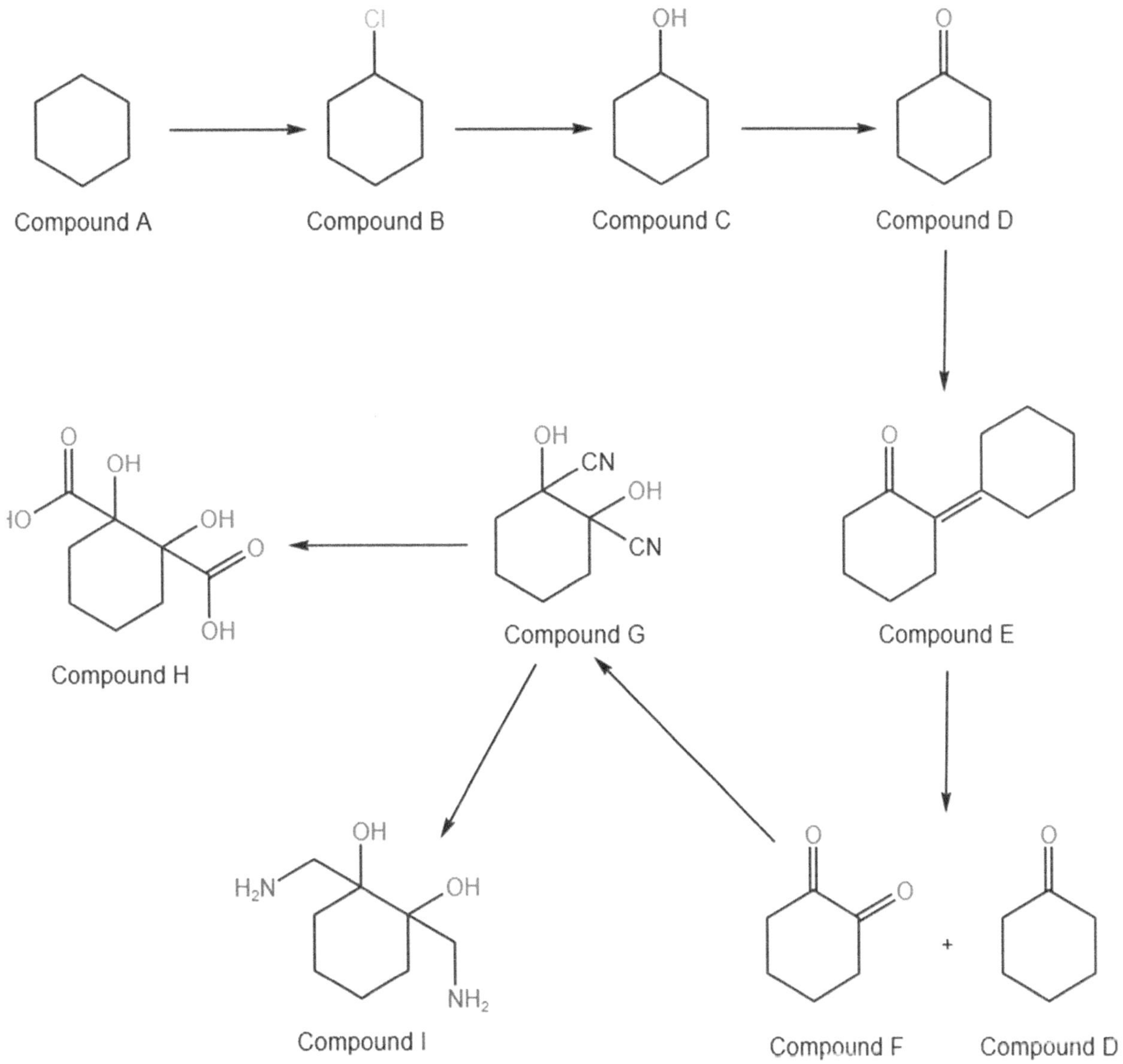

Base-catalyzed Aldol condensation Mechanism (FYI)

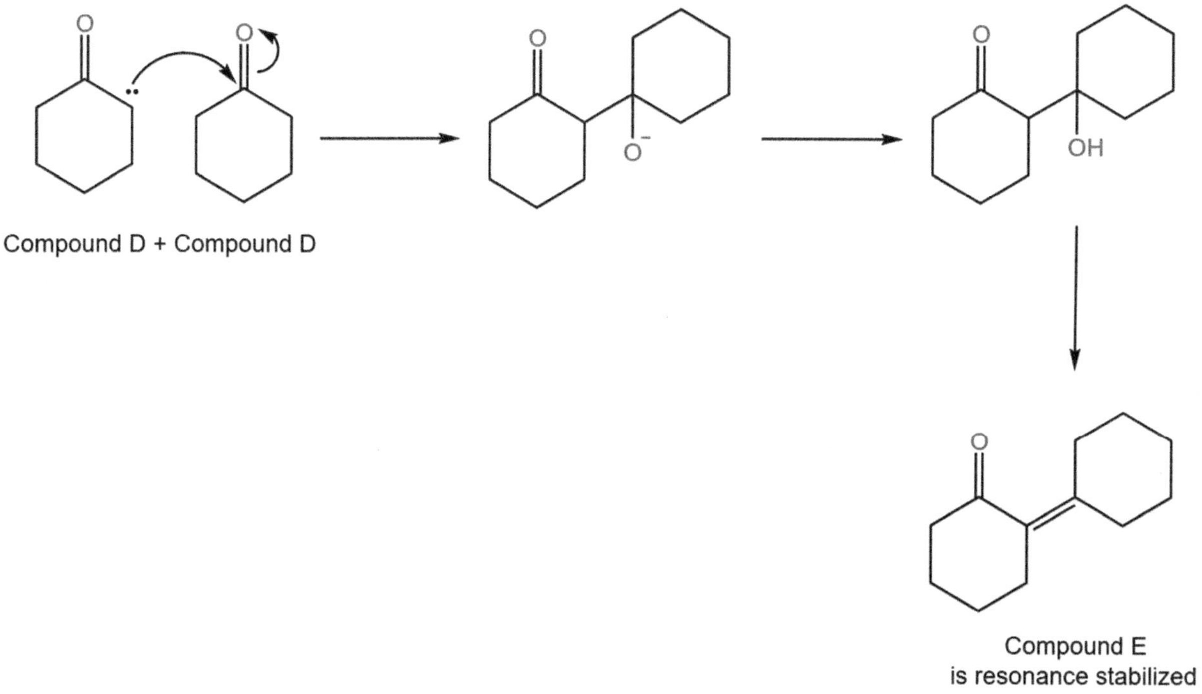

Question 81:

Compound **A** (C_8H_{10}) exhibits two distinct peaks in its ^{1}H NMR spectrum, indicating that its hydrogen atoms are distributed in two chemically different environments. Reaction of **A** with potassium manganate(VII) results in the formation of **B** ($C_8H_6O_4$).

Treatment of **A** with ethanoyl chloride in the presence of anhydrous aluminum chloride produces **C** ($C_{10}H_{12}O$). Analysis of **C** using IR spectroscopy reveals a strong absorption band at 1685 cm^{-1}. Reduction of **C** with sodium borohydride in pyridine yields **D** ($C_{10}H_{14}O$), which exists in stereoisomeric forms. IR spectroscopy of **D** shows a broad absorption band at 3340 cm^{-1}.

Heating **D** with concentrated sulfuric acid results in the elimination of a molecule, producing **E**, which displays an absorption band near 1630 cm^{-1} in its IR spectrum. When **C** is reacted in the presence of sodium ethanoate, followed by acidification, two molecules of **C** combine with the loss of a water molecule to form **F** ($C_{20}H_{22}O$).

Treating **F** with bromoethane and magnesium in dry ether results in the formation of **G** ($C_{22}H_{28}O$). Additionally, bubbling hydrogen cyanide gas in the presence of a catalytic amount of NaOH through **F** leads to the generation of **H** ($C_{21}H_{23}NO$).

Your task is to identify compounds **A** through **H**, providing detailed explanations for the observations recorded during each transformation. Ensure to support your reasoning with balanced chemical equations and logical deductions.

My thought process:

Compound **A** is a C_8H_{10} hydrocarbon with four degrees of unsaturation—one benzene ring plus two extra carbons in substituents. Its ^{1}H NMR shows just two signals, which tells us all the ring protons are equivalent and all the alkyl protons are equivalent. The only structure that fits is 1,4-dimethylbenzene.

When 1,4-dimethylbenzene is oxidised with hot $KMnO_4$, each methyl side-chain is converted into a –COOH, giving benzene-1,4-dicarboxylic acid ($C_8H_6O_4$), compound **B**.

Separately, treating **A** with ethanoyl chloride in the presence of anhydrous $AlCl_3$ effects a Friedel–Crafts acylation at the ring. Because both methyl groups in 1,4-dimethylbenzene are ortho/para directors and are symmetrically placed, only one mono-acylated product forms: 1-ethanoyl-2,5-dimethylbenzene ($C_{10}H_{12}O$), compound **C**. Its strong IR absorption at 1685 cm^{-1} confirms the newly introduced ketone.

Reduction of **C** with sodium borohydride in pyridine selectively converts the ketone to a secondary alcohol, yielding compound **D**, which exists as a pair of enantiomeric alcohols. A broad O–H stretch at 3340 cm^{-1} in the IR spectrum attests to the formation of a hydrogen-bonded –OH group.

Heating **D** with concentrated H_2SO_4 causes dehydration of the secondary (benzylic) alcohol. The resulting alkene, compound **E**, shows a C=C band near 1630 cm^{-1} in its IR spectrum.

When two molecules of **C** are treated with sodium ethanoate and then acidified, they undergo an aldol-type condensation: first coupling to form a β-hydroxy ketone, then eliminating water to give compound **F**, an α,β-unsaturated ketone (also called enone) with formula $C_{20}H_{22}O$. Its ten degrees of unsaturation (two rings + one C=O + one C=C) align perfectly with that structure.

When **F** is treated with ethylmagnesium bromide—generated from bromoethane and Mg in dry ether—it undergoes a direct 1,2-nucleophilic addition to form compound **G**. In contrast, bubbling HCN through **F** in the presence of a catalytic amount of NaOH induces a conjugate (1,4-) Michael addition of cyanide across the enone, yielding the cyano-substituted product **H** ($C_{21}H_{23}NO$).

Why this problem matters:
It brings together key A-level transformations—side-chain oxidation, Friedel–Crafts acylation, selective reduction, dehydration, aldol condensation, Grignard addition, and conjugate addition—along with spectroscopic diagnosis (IR and NMR). Mastery of each step and the ability to read subtle clues in spectroscopy will make questions like this routine rather than daunting.

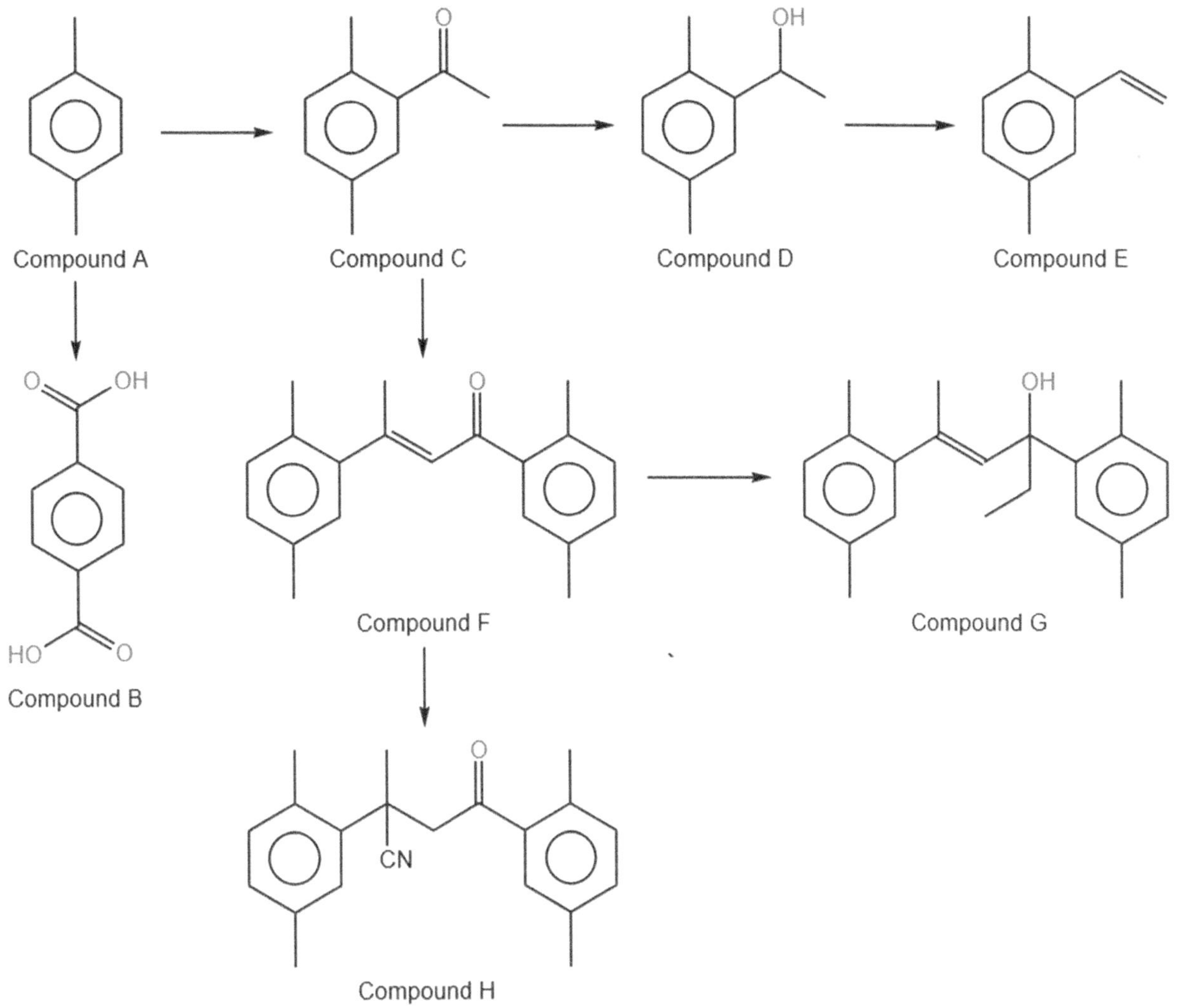
O
OH
Compound A
Compound C
Compound D
Compound E
O
OH
HO
O
Compound B
O
Compound F
OH
Compound G
O
CN
Compound H

Final Reflections: Beyond the Last Question

"To be understood is a rarefied triumph; yet even in his insight, there lingered an abyss—a chasm where understanding dances tantalizingly close, yet remains forever elusive." — **DT**

I hope this book has made the world of **structure elucidation** a little less intimidating — and perhaps even a little more enjoyable. Whether you read it from start to finish or dipped in and out for guidance, I trust it has served as a helpful companion on your journey through organic chemistry.

In the very first chapter, I opened with a bold statement:

"Organic Chemistry Isn't Complicated — It Just Pretends to Be."

Looking back now, I find myself revisiting that claim. Perhaps organic chemistry *is* complicated — not because it's trying to confuse us, but because it reflects a world of layered logic, shifting patterns, and elegant exceptions. **Complete understanding is a noble pursuit**, but like the best ideas in science, it's also a **moving target**.

The **Heisenberg Uncertainty Principle** offers a useful metaphor here. The deeper we look, the more we alter what we observe. Learning works the same way: the more we uncover, the more we realise how much lies beyond our reach. But that's not a flaw — it's part of the beauty. Chemistry, like life, is full of **uncertainties that inspire curiosity** rather than fear.

There's meaning in the number of questions in this book: **81**. This is drawn from the Chinese philosophies of **"九九八十一难"** — representing **81 tribulations** on the path to transformation — and **"九九归真"**, a return to truth and simplicity through lived experience. These ancient ideas honour the journey of **growth through struggle**, and I wanted this book to reflect that same spirit.

Each question wasn't just designed to test you — it was meant to **transform you**. Like the mythical **phoenix**, we rise stronger with every difficulty we overcome. These problems are more than just exercises. They are opportunities — to learn, to reflect, and to see the deeper elegance embedded in the logic of organic chemistry.

As you continue your learning, remember this: **mastery is not a destination**. It's the art of being adaptable, of letting what you've learned flow through you with confidence and clarity. Your understanding may never feel "complete" — and that's exactly what keeps learning alive. What matters most is your willingness to keep **exploring**, **questioning**, and **growing**.

Leonard and I have poured our combined experience into this book, and we hope it has guided you well.

Thank you for allowing me to be part of your journey.
The rest, as always, is in your hands.

— **David Tan & Leonard Pereira**

INDEX

Printed in France by Amazon
Brétigny-sur-Orge, FR

50148400R00167